Power System
SCADA
and
Smart Grids

Power System
SCADA
and
Smart Grids

Mini S. Thomas
Jamia Millia Islamia University
New Delhi, India

John D. McDonald
GE Energy Management - Digital Energy
Atlanta, Georgia, USA

CRC Press
Taylor & Francis Group
Boca Raton   London   New York

CRC Press is an imprint of the
Taylor & Francis Group, an **informa** business

CRC Press
Taylor & Francis Group
6000 Broken Sound Parkway NW, Suite 300
Boca Raton, FL 33487-2742

Printed on acid-free paper
Version Date: 20150203

International Standard Book Number-13: 978-1-4822-2674-4 (Hardback)

Visit the Taylor & Francis Web site at
http://www.taylorandfrancis.com

and the CRC Press Web site at
http://www.crcpress.com

Contents

Preface

Although SCADA systems have revolutionized the way complex, geographically distributed industrial systems are monitored and controlled, the details about SCADA components, implementations and application functions have largely remained proprietary. Engineers learn the fundamentals of this evolving technology, mostly on the job and students have difficulty in gathering information as the literature on SCADA fundamentals is scarce and scattered. With the smart grid initiatives taking a giant leap in recent times, it is imperative to have sound knowledge of SCADA basics to implement all the functionalities effectively.

Hence this book is an attempt to bring the fundamentals of SCADA systems and elaborate the possible application functions so that academia and practitioners stand to gain from the content. The book is dedicated to power system SCADA, although SCADA systems are extensively used in other industrial sectors like oil and gas, water supply, etc. The discussion revolves around SCADA fundamentals in the initial chapters, followed by application functions from generation, transmission, distribution, and customer automation functions. The first chapter provides an overview of SCADA systems, evolution, and use of SCADA in power systems, the power system field, and the data acquisition process. Chapter two is the soul of the book where the building blocks of SCADA systems are discussed in detail from the legacy Remote Terminal Units (RTUs) to the latest Intelligent Electronic Devices (IEDs), data concentrators, and master stations. The building of different SCADA systems is elaborated with practical implementation descriptions.

Communications is of utmost importance in power system SCADA as the field is widely distributed over a large geographical area and owing to the time bound data transmission requirements in milliseconds. Chapter three gives a comprehensive discussion of the data communication, protocols, and media usage. Chapter four discusses substation automation which forms the basis for transmission, distribution, and customer automation. Chapter five discusses energy management systems for transmission control centers with specific emphasis on generation operation and management, real-time transmission operation and management and study mode simulations. Distribution automation and distribution management systems (DMS) are discussed in detail in Chapter six with real time, advanced analytical DMS functionalities, and DMS integration with other distribution application. Chapter seven introduces readers to smart grid concepts discussing the building blocks of smart distribution and smart transmission.

The book is intended to catch the attention of practitioners, fresh and experienced alike, to acquire basic knowledge of SCADA systems and application functions, which are evolving day by day, to help them adapt to the new challenges effortlessly. Senior undergraduate and graduate students will find the content very useful with the description of each and every component of SCADA systems and the application functionalities.

This book is the outcome of a dream to assist academia and industry in enhancing the understanding of SCADA systems which has fascinated both of us over the years. With one person gaining experience from dedicating his entire career in designing and implementing new SCADA systems across the world for utilities and the other, learning and developing SCADA laboratory systems for students to learn and experiment, it was a natural choice to pen down the content for the users of SCADA systems.

However, this journey was not possible without the help of a few friends who believed in us and helped us with motivation, support and suggestions. We thank Nora Konopka, of CRC Press who put her trust in us and helped us write this book with her gentle push once in a while. Professor Saifur Rahman, Director, ARI, Virginia Tech has always been a supporter and thanks are due to Dr. Jiyuan Fan, who helped us finalize the content of the book. The support received from faculty and students of Jamia Millia Islamia, especially Anupama Prakash, Ankur Singh Rana, Namrata Bhaskar and Praveen Bansal, is gratefully appreciated. The support received from our families, especially Shaji and Jo-Ann, our spouses, Shobha & Mathew and Sarah & Mark, our children, is acknowledged.

We do hope that this book will bring a better understanding of the inner secrets of SCADA systems, unveil the potential of the smart grid and inspire more minds to get involved.

Mini S. Thomas
John D. McDonald

The authors

Mini S. Thomas is a professor in the Department of Electrical Engineering, Faculty of Engineering and Technology, Jamia Millia Islamia (JMI), and has 29 years of teaching and research experience in the field of power systems. Currently she is the Director of Centre for Innovation and Entrepreneurship at the University. She was the head of the Department of Electrical Engineering from 2005 to 2008. Thomas was a faculty member at Delhi College of Engineering, Delhi (now DTU), and at the Regional Engineering College (now NIT) Calicut, Kerala, before joining Jamia. She grad-uated from the University of Kerala (Gold Medalist), and completed her MTech from IIT Madras (Gold Medalist, Siemens prize) and PhD from IIT Delhi, India, all in electrical engineering.

Thomas has done extensive research work in the areas of supervisory control and data acquisition (SCADA) systems, substation and distribution automation, and smart grid. She has published over 100 research papers in international journals and conferences of repute, has successfully completed many research projects, and is the coordinator of the special assistance program (SAP) on power system automation from UGC, Government of India. She is also a reviewer of prominent journals in her field.

The first SCADA laboratory and substation automation (SA) laboratory were set up by Thomas at JMI, and they are bringing laurels to the university in terms of research publications, memoranda of understanding (MOUs), and training opportunities, and more importantly, an enhanced image among the world power engineering fraternity. She, as the founder coordinator, with industry participation, has drafted the curriculum and started a unique, first full-time MTech program in the

Faculty of Engineering and Technology at JMI in 2003 in electrical power system management, which offers unique courses on power automation and novel hands-on training.

Thomas has worked continuously for industry and academia interaction; the MTech program and the SCADA and SA laboratories have been set up with industry collaboration. She was instrumental in signing an MOU with Power Grid Corporation of India Limited (PGCIL), the transmission utility of India, for long-term cooperation with JMI. She and her team regularly conduct training and certification programs for control center operators of Power System Operations Corporation (POSCOCO) in SCADA basics. She is a certified trainer for "Capacity Building of Women Managers in Higher Education" by UGC and has conducted many training sessions for empowerment. She initiated the Center for Innovation and Entrepreneurship at JMI to promote innovation and business development for students and faculty members.

She received the Career Award for young teachers from the government of India, and has won the IEEE MGA Larry K Wilson transnational award, MGA Innovation award, Outstanding Volunteer award, Outstanding Branch Counselor award, and Power and Energy Society (PES) Outstanding Chapter Engineer award, to name a few.

Thomas is very active in professional societies and has served on the global boards of IEEE. She is currently a member of the PES LRP (long-range planning) committee. She was a board member of IEEE PSPB (publication services and products board), educational activities board (EAB), served as the vice chair of IEEE MGA (member and geographic activities) board, and was the Asia Pacific student activities coordinator. She has experience of over a decade on international boards and committees of the IEEE and is currently IEEE Delhi section chairperson.

She has traveled extensively around the globe, delivered lectures at prestigious universities, and has interacted with technical experts all over the world.

John D. McDonald, P.E., is director of Technical Strategy and Policy Development for GE Energy Management's Digital Energy business. He has 40 years of experience in the electric utility industry. McDonald joined GE in 2008 as general manager, marketing, for GE Energy's Transmission and Distribution (now Digital Energy) business. In 2010, he accepted his current role of director, Technical Strategy and Policy Development and is responsible for setting and driving the vision that integrates GE's standards participation, and Digital Energy's industry organization participation through leadership activities, regulatory/policy participation, education programs, and product/systems development for designing comprehensive solutions for customers.

McDonald is a sought-after industry leader, technical expert, educator, and speaker. In his 28 years of working group and subcommittee leadership with the IEEE Power and Energy Society (PES) Substations Committee, he led seven working groups and task forces that published standards and tutorials in the areas of distribution SCADA, master and remote terminal unit (RTU), and RTU/IED communications protocols. He was elected to the board of governors of the IEEE-SA (standards association) for 2010 to 2011, focusing on long-term IEEE smart grid standards strategy. McDonald was elected to chair the NIST Smart Grid Interoperability Panel (SGIP) Governing Board from 2010 to 2012. He is presently chairman of the board for SGIP 2.0, Inc., the member-funded nonprofit organization.

He is past president of the IEEE PES, chair of the Smart Grid Consumer Collaborative (SGCC) board, member of the IEEE PES Region 3 Scholarship Committee, the vice president for technical activities for the US National Committee (USNC) of CIGRE, and the past chair of the IEEE PES Substations Committee. He was the IEEE Division VII director in 2008 to 2009. McDonald is a member of the advisory committee for the annual DistribuTECH Conference, vice chair of the Texas A&M University Smart Grid Center advisory board, and member of the Purdue University Office of Global Affairs Strategic Advisory Council. He received the 2009 Outstanding Electrical and Computer Engineer Award from Purdue University.

McDonald teaches a smart grid course at the Georgia Institute of Technology, a smart grid course for GE, and substation automation, distribution SCADA, and communications courses for various IEEE PES local chapters as an IEEE PES distinguished lecturer. He has published 60 papers and articles in the areas of SCADA, SCADA/EMS, SCADA/DMS, and communications, and is a registered Professional Engineer (Electrical) in California and Georgia.

He received his BSEE and MSEE (power engineering) degrees from Purdue University, and an MBA (finance) degree from the University of California-Berkeley. He is a member of Eta Kappa Nu (Electrical Engineering Honorary) and Tau Beta Pi (Engineering Honorary), a fellow of IEEE, and was awarded the IEEE Millennium Medal in 2000, the IEEE PES Excellence in Power Distribution Engineering Award in 2002, and the IEEE PES Substations Committee Distinguished Service Award in 2003.

McDonald was editor of the substations chapter, and a co-author, of *The Electric Power Engineering Handbook* (co-sponsored by the IEEE PES and published by CRC Press, Boca Raton, FL, 2000). He was also editor-in-chief for *Electric Power Substations Engineering* (3rd ed., Taylor & Francis/ CRC Press, 2012).

chapter one

Power system automation

1.1 Introduction

The global electricity demand is growing at a rapid pace, making the requirements for more reliable, environment friendly, and efficient transmission and distribution systems inevitable. The traditional grids and substations are no longer acceptable for sustainable development and environment-friendly power delivery. Hence, the utilities are moving toward the next-generation grid incorporating the innovations in diverse fields of technology, thereby enabling the end users to have more flexible choices and also empowering the utilities to reduce peak demand and carbon dioxide emissions to become more efficient in all respects.

Power engineering today is an amalgam of the latest techniques in signal processing, wide area networks, data communication, and advanced computer applications. The advances in instrumentation, intelligent electronic devices (IEDs), Ethernet-based communication media coupled with the availability of less-expensive automation products and standardization of communication protocols led to the widespread automation of power systems, especially in the transmission and distribution sector.

In today's world with limited resources and increasing energy needs, optimization of the available resources is absolutely essential. Conventional power generation resources such as coal, water, and nuclear fuels are either depleting or raising environmental concerns. Renewable sources are also to be utilized judiciously. Hence there is a need to optimize the energy use and reduce waste. Automation of power systems is a solution toward this goal, and every sector of the power system, from generation, to transmission to distribution to the customer is being automated today to achieve optimal use of energy and resources.

In order to integrate the new technologies with the existing system, it is necessary that the practicing engineers are well versed with the old and new technologies. However, in the present scenario, most of the engineering professionals learn the new technology "on the job" as the pace of technology development is very fast with the advent of new communication protocols, relay IEDs, and related functions. This is all the more relevant in the core field of power engineering as the power industry needs trained engineers to keep up the pace of the rapid expansion the power industry is envisaging, to meet the energy consumption that is expected to triple by 2050. It is pertinent to explore the automation of power systems in detail.

1.2   Evolution of automation systems

The evolution of automation systems could be traced back to the first industrial revolution (1750–1850), when the work done by the human muscle was replaced by the power of machines. During the second industrial revolution (1850–1920), process control was introduced and the routine functions of the human mind and continuous presence were taken over by machines. The human mind was relieved of the bulk and tedious physical and mental activities. Michael Faraday invented the electric motor in 1821, and James Clark Maxwell linked electricity and magnetism in 1861–1862. In the later part of the nineteenth century, there were rapid developments in electricity and supply of electric power with the giants like Siemens, Westinghouse, Nikola Tesla, Alexander Graham Bell, Lord Kelvin, and many others contributing immensely. In 1891, the first long-distance three-phase transmission line of high power was featured at the International Electro-Technical Exhibition in Frankfurt. Along with the developments in electric power generation, transmission, and distribution to customers, the automation including remote monitoring and control of electric systems became inevitable.

The initial control equipment consisted of analog devices which were large and bulky, and the control rooms had huge panels with innumerable wires running from the field to the control center. The operator could not make use of the information available, as during an emergency, a number of events occurred simultaneously and it was impossible to handle all of them since there was no intelligent alarm processing. Excessive cost was associated with a reconfiguration or expansion of the system. The expensive space requirement was also a constraint in the case of analog control, as the control panels were large. Storage of information was also an issue, as for power systems post-event analysis is crucial.

With the introduction of computers into the automation scenario, automation became more operator friendly, although initially computer use was restricted to data storage and to change set points for analog controllers. Early digital computers had serious disadvantages such as minimal memory, poor reliability, and programming written in machine language.

Two major developments led to the advent of distributed control: the advances in integrated circuits and in communication systems. Distributed control systems were modular in structure, with preprogrammed menus, having a wide selection of control algorithms for execution. The data highway became possible with the introduction of new communication techniques and media. Redundancy at any level was possible, due to the availability of components at cheaper rates, and extensive diagnostic tools became part of the supervisory control and data acquisition (SCADA) systems.

1.2.1   History of automation systems

Supervisory control and data acquisition (SCADA) systems are widely used for automation of the power sector and represent an evolving field, with new products and services added on a daily basis. Detailed study of SCADA systems is essential for power automation personnel to understand the integration of devices, to understand the communication between components, and for proper monitoring and control of the system in general.

There were undoubtedly many methods of remote control invented by early pioneers in the supervisory control field which have long since been forgotten. Control probably began with an operator reading a measurement and taking some mechanical control action as a result of that measurement.

Most early patents on supervisory control were issued between 1890 and 1930. These patents were granted mainly to engineers working for telephone and other communication industries. Almost all patents involving remote control closely followed the techniques of the first automatic telephone exchange installed in 1892 by Automatic Electric Company.

From 1900 until the early 1920s many varieties of remote control systems were developed. Most of these, however, were of only one class or the other (i.e., either remote control or remote supervision [monitoring only]). One of the earliest forerunners of the modern SCADA system was a system designed in 1921 by John B. Harlow. Harlow's system automatically detected a change of status at a remote station and reported this change to a control center. In 1923, John J. Bellamy and Rodney G. Richardson developed a remote control system employing an equivalent of our modern "check-before-operate" technique to ensure the validity of a selected control point before the actual control was initiated. The operator could also ask for a point "check" to verify its status.

The first logging system was designed by Harry E. Hersey in 1927. This system monitored information from a remote location and printed any change in the status of the equipment together with the reported time and date when the change took place.

As the scope of supervisory control applications changed, so did many of the fundamentals of supervisory control technology. During the early years all of the systems were electromechanical. The supervisory systems evolved to using solid-state components, electronic sensors, and analog-to-digital convertors. In this evolution, however, the same remote terminal unit (RTU) configuration was maintained. The companies making the RTUs merely upgraded their technology without looking at alternate ways of performing the RTU functions. In the 1980s process control companies began applying their technology and technical approach to the

SCADA electric utility market. As a result, RTUs used microprocessor-based logic to perform expanded functions. The application of micropro-cessors increased the flexibility of supervisory systems and created new possibilities in both operation and capabilities.

1.3   Supervisory control and data acquisition (SCADA) systems

Automation is used worldwide in a variety of applications ranging from the gas and petroleum industry, power system automation, building automation, to small manufacturing unit automation. The terminology *SCADA* is generally used when the process to be controlled is spread over a wide geographic area, like power systems. SCADA systems, though used extensively by many industries, are undergoing drastic changes. The addition of new technologies and devices poses a serious challenge to educators, researchers, and practicing engineers to catch up with the latest developments.

SCADA systems are defined as a collection of equipment that will provide an operator at a remote location with sufficient information to determine the status of particular equipment or a process and cause actions to take place regarding that equipment or process without being physically present.

SCADA implementation thus involves two major activities: data acquisition (monitoring) of a process or equipment and the supervisory control of the process, thus leading to complete automation. The complete automation of a process can be achieved by automating the *monitoring* and the *control* actions.

Automating the monitoring part translates into an operator in a con-trol room, being able to "see" the remote process on the operator console, complete with all the information required displayed and updated at the appropriate time intervals. This will involve the following steps:

- Collect the data from the field.
- Convert the data into transmittable form.
- Bundle the data into packets.
- Transmit the packets of data over the communication media.
- Receive the data at the control center.
- Decode the data.
- Display the data at the appropriate points on the display screens of the operator.

Automating the control process will ensure that the control command issued by the system operator gets translated into the appropriate action in the field and will involve the following steps:

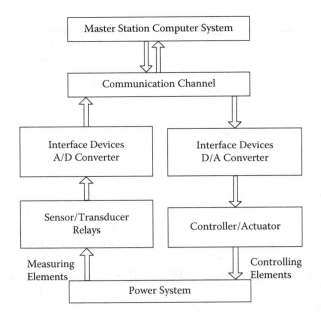

Figure 1.1 The monitoring and controlling process.

- The operator initiates the control command.
- Bundle the control command as a data packet.
- Transmit the packet over the communication media.
- The field device receives and decodes the control command.
- Control action is initiated in the field using the appropriate device actuation.

The set of equipment *measuring elements* helps in acquiring the data from the field, and the set of equipment *controlling elements* implements the control commands in the field, as shown in Figure 1.1.

1.3.1   Components of SCADA systems

SCADA is an integrated technology composed of the following four major components:

1. *RTU*: RTU serves as the eyes, ears, and hands of a SCADA system. The RTU acquires all the field data from different field devices, as the human eyes and ears monitor the surroundings, process the data and transmit the relevant data to the master station. At the same time, it distributes the control signals received from the master station to the field devices, as the human hand executes instructions from the brain. Today Intelligent Electronic Devices (IEDs) are replacing RTUs.

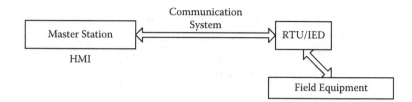

Figure 1.2 Components of SCADA systems.

2. *Communication System*: This refers to the communication channels employed between the field equipment and the master station. The bandwidth of the channel limits the speed of communication.
3. *Master Station*: This is a collection of computers, peripherals, and appropriate input and output (I/O) systems that enable the operators to monitor the state of the power system (or a process) and control it.
4. *Human-Machine Interface (HMI)*: HMI refers to the interface required for the interaction between the master station and the operators or users of the SCADA system.

Figure 1.2 illustrates the components of a SCADA system.

1.3.2   SCADA applications

SCADA systems are extensively used in a large number of industries, for their monitoring and control. The oil and gas industry uses SCADA extensively for the oil fields, refineries, and pumping stations. The large oil pipelines and gas pipelines running across the oceans and continents are also monitored by appropriate SCADA systems, where the flow, pressure, temperature, leak, and other essential features are assessed and controlled. Water treatment, water distribution, and wastewater management systems use SCADA to monitor and control tank levels, remote and lift station pumps, and the chemical processes involved. SCADA systems control the heating, ventilation, and air conditioning of buildings such as airports and large communication facilities. Steel, plastic, paper, and other major manufacturing industries utilize the potential of SCADA systems to achieve more standardized and quality products. The mining industry with integrated SCADA for the mining processes, like tunneling, product flow optimization, material logistics, worker tracking, and security features, is the latest addition to the list, making *digital mines*.

The use of SCADA systems in the power industry is widespread, and the rest of the discussion in this chapter will focus specifically on the power sector, including generation, transmission, and distribution of power.

1.4   SCADA in power systems

SCADA systems are in use in all spheres of power system operations starting from generation, to transmission, to distribution, and to utilization of electrical energy. The SCADA functions can be classified as basic and advanced application functions.

1.4.1   SCADA basic functions

The basic SCADA functions include data acquisition, remote control, human-machine interface, historical data analysis, and report writing, which are common to generation, transmission, and distribution systems.

Data acquisition is the function by which all kinds of data—analog, digital, and pulse—are acquired from the power system. This is accomplished by the use of sensors, transducers, and status point information acquired from the field.

Remote control involves the control of all the required variables by the operator from the control room. In power systems, the control is mostly of switch positions; hence, digital control output points are abundant, such as circuit breaker and isolator positions and equipment on and off positions.

Historical data analysis is an important function performed by the power system SCADA, where the post-event analysis is done using the data available after the event has happened. An example is the post-outage analysis where the data acquired by the SCADA system can provide insights into such information as the sequence of events during the outage, malfunctioning of any device in the system, and the action taken by the operator. This could be a powerful tool for future planning and is extensively used by power engineering personnel.

Power system SCADA requires a number of reports to be generated for consumption at different levels of the management and from different departments of the utility. Hence, report generation is essential as per the requirements of the parties and departments involved.

1.4.2   SCADA application functions

Figure 1.3 illustrates the use of SCADA in power systems, with the initial SCADA block depicting the basic functions, as discussed in Section 1.4.1. The right section of the figure illustrates the generation SCADA, represented by SCADA/AGC (Automatic Generation Control), implemented in the generation control centers across the world. Further, the transmission SCADA is shown as SCADA/EMS (Energy Management Systems) where the basic functions are supplemented by the energy management system functions. This is implemented in the transmission control centers.

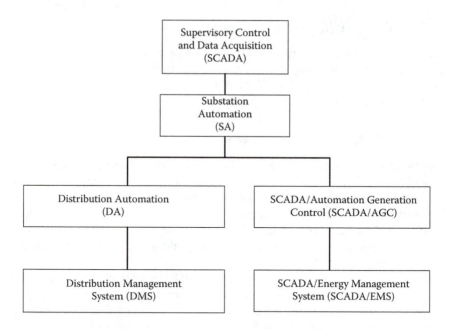

Figure 1.3 Use of SCADA in power systems.

The EMS software applications are the most expensive component of the SCADA/EMS, mainly due to the complexity of each application. The left part of the figure shows the distribution functions superimposed on the basic SCADA functions, beginning at the SCADA/distribution automation system and further expanding to the distribution management system functions. As one scans the figure from top to bottom, the systems become more complex and more expensive (i.e., the basic SCADA system is the simplest and least expensive, the SCADA/AGC is more involved and a little more expensive, and the SCADA/EMS is much more complex and expensive). The same is true for distribution. The SCADA/DA is more involved and more expensive than the basic SCADA system. The SCADA/DMS is much more complex and expensive.

1.4.2.1 Generation SCADA application functions

As discussed earlier, generation SCADA, in addition to the basic functions discussed earlier, will include the following application functions.

- *Automatic Generation Control (AGC)*: a compendium of equipment and computer programs implementing closed-loop feedback control of frequency and net interchange
- *Economic Dispatch Calculation (EDC)*: the scheduling of power from all available sources in such a way to minimize cost within some security limit

- *Interchange Transaction Scheduling (ITS)*: ensures that sufficient energy and capacity are available to satisfy load energy and capacity requirements
- *Transaction Evaluation (TE)*: evaluates economy of transactions using the unit commitment results as the base condition
- *Unit Commitment (UC)*: produces the hourly start-up and loading schedule which minimizes the production cost for up to one week in the future
- *Short-Term Load Forecasting (STLF)*: produces the hourly system load for up to one week into the future and is used as input to the unit commitment program
- *Hydrothermal coordination*: the scheduling of power from all available hydro generation in such a way to minimize cost within constraints (e.g., reservoir levels)

1.4.2.2 Transmission SCADA application functions
The transmission SCADA will include energy management system (EMS) functions such as

- *Network Configuration/Topology Processor*: analyzes the status of circuit breakers as well as measurements to automatically determine the current model of the power system
- *State Estimation*: provides a means of processing a set of redundant information to obtain an estimate of the state variables of the system
- *Contingency Analysis*: simulates outages of generating units and transmission facilities to study their effect on bus voltages, power flows, and the transient stability of the power system as a whole
- *Three-Phase Balanced Power Flow*: obtains complete voltage angle and magnitude information for each bus in a power system for specified load and generator real power and voltage conditions
- *Optimal Power Flow*: optimize some system objective function, such as production cost, losses, and so on, subject to physical constraints on facilities and the observation of the network laws

Details of the above functions and additional functions are explained in Chapter 5.

1.4.2.3 Distribution automation application functions
Distribution automation/distribution management systems (DA/DMS) include substation automation, feeder automation, and customer automation. The additional features incorporated in distribution automation will be

- Fault identification, isolation, and service restoration
- Network reconfiguration

- Load management/demand response
- Active and reactive power control
- Power factor control
- Short-term load forecasting
- Three-phase unbalanced power flow
- Interface to customer information systems (CISs)
- Interface to geographical information systems (GISs)
- Trouble call management and interface to outage management systems (OMSs)

Details of distribution automation functions are given in Chapter 6.

1.5   Advantages of SCADA in power systems

Automating a system brings many advantages, and the case of power systems is no different. Some of the advantages are as follows:

- Increased reliability, as the system can be operated with less severe contingencies and the outages are addressed quickly
- Lower operating costs, as there is less personnel involvement due to automation
- Faster restoration of power in case of a breakdown, as the faults can be detected faster and action taken
- Better active and reactive power management, as the values are accurately captured in the automation system and appropriate action can be taken
- Reduced maintenance cost, as the maintenance can be more effectively done (transition from time-based to condition-based maintenance) with continuous monitoring of the equipment
- Reduced human influence and errors, as the values are accessed automatically, and the meter reading and related errors are avoided
- Faster decision making, as a wealth of information is made available to the operator about the system conditions to assist the operator in making accurate and appropriate decisions
- Optimized system operation, as optimization algorithms can be run and appropriate performance parameters chosen

Some of the additional benefits by SCADA system implementation are as discussed below.

1.5.1   Deferred capital expenditure

With a real-time view of loading on various transmission lines, feeders, transformers, circuit breakers, and other equipment, and the ability to control from a central location, utilities can achieve proper load balancing on the system, avoiding unnecessary overloading of equipment and ensuring a longer service life for the components. Better equipment monitoring and load balancing can extend the economic life of the primary equipment and thus defer certain capital expenditures on assets. More capacity can be squeezed out of the existing equipment with proper monitoring; hence, additional expansion can be deferred for a while when the load increases.

1.5.2   Optimized operation and maintenance costs

Utilities can achieve significant savings in operation and maintenance (O&M) costs through the SCADA implementation. Functions such as predictive maintenance, volt-var control, self-diagnostic programs, and access to the automation data help the utility to optimize their costs by allowing it to make better informed decisions on O&M strategies that are based on comprehensive and accurate operational data rather than rules of thumb.

1.5.3   Equipment condition monitoring (ECM)

By implementing equipment condition monitoring, vital equipment parameters are automatically tracked to detect abnormalities, and with proper maintenance and care, the life of costly equipment can be extended. Intelligent electronic devices (IEDs) are available which continuously monitor equipment health. This helps in resolving issues in the preliminary stages rather than later, and hence major equipment failure and service disruption can be avoided. Power transformers, bushings, tap changers, and substation batteries are some of the power system equipment monitored by ECM IEDs.

1.5.4   Sequence of events (SOE) recording

Crucial events in the system are time stamped for post-event analysis, and this provides crucial data about the system loading patterns. Later it is easy to recreate events in the same sequence as they happened in the system which goes a long way in helping planners design the new transmission lines, feeders, and networks for the future.

1.5.5   Power quality improvement

Power quality (PQ) monitoring devices can be connected to the network and monitored centrally to monitor harmonics, voltage sags, swells, and unbalances. Corrective measures such as switching of capacitor banks and voltage regulators can be implemented to improve the power quality, so that the customer receives quality power supply all the time.

1.5.6   Data warehousing for power utilities

The introduction of IEDs and availability of high-speed communication systems have made it possible to convey the operational data to the SCADA master station and the nonoperational data, including the digitized wave-forms, to the enterprise data warehouse. The data are archived, and the analysis of the data is driven by the urge to provide more reliable supply to the customer and to make the system operation more competitive. The major benefits of the data analysis include explaining why systems behave abnormally, restoring outages faster, preventing problems from escalating, operating equipment more efficiently, making informed decisions about infrastructure repair and replacement, keeping equipment healthy and extending equipment life, improving reliability and availability, maximizing the utilization of existing assets, improving employee efficiency, and increasing profitability. The major user groups that have been identified in a power utility which will benefit from data warehousing are the operations department, planning department, protection department, engineering department, maintenance department, asset management department, power quality department, purchasing department, marketing department, safety department, and customer support department.

Thus, automation brings in a new set of solutions for better managing the assets for customer satisfaction and reliable operation of the system. Hence, utilities across the world are embracing SCADA systems and are reaping the associated benefits.

1.6   Power system field

Electricity is generated at the generating stations and transmitted over the transmission system to the distribution substation, from where it is distributed to the consumers. In the current scenario, to this traditional system, renewable generation is added at transmission and distribution, including the customer premises. Hence, SCADA systems will acquire data from all these components, and a brief discussion of these components follows.

1.6.1   Transmission and distribution systems

The generated electricity reaches the customer premises passing through a variety of substations which are classified as follows:

- Switchyard or generating substations
- Bulk power substations or grid substations

- Distribution substations
- Special-purpose substations (e.g., traction substation, mining substation, mobile substations, etc.)

A transmission substation (generating or grid substation) usually has the following components:

- Transformers (with or without tap changers)
- Station buses and insulators
- Current transformers
- Potential transformers
- Circuit breakers
- Disconnecting switches (isolators or fuses)
- Reactors, series or shunt
- Capacitors, series or shunt
- Relays/relay IEDs
- Substation batteries
- Line or wave trap and coupling capacitors for power line carrier communication

The present-day distribution substations have similar equipment with reactive and capacitive compensation equipment in place, however with all the equipment at a lower rating. Figure 1.4 shows a typical substation.

As far as the SCADA systems are concerned, analog data are acquired from the transformers and station buses via the current and voltage transducers and are further processed for transmission to the control room. Status data (digital data) are acquired from the circuit breakers, isolators, and the shunt and series compensation devices (on/off positions) and are conveyed to the master station as per the requirement. Environmental data such as temperature, pressure, humidity, and weather conditions are

Figure 1.4 A typical substation.

collected by the appropriate sensors and are processed for onward transmission to the control center.

1.6.2   Customer premises

With the customer taking center stage in an automated distribution system, the devices in the customer premises hold the key to successful implementation of the future smart grid. The smart energy meter capable of two-way communication, the smart appliances in the house, and also the smart plugs with communication facility are inevitable for customer automation. The main challenge here will be the integration of existing plugs and devices with the new smart meters at the customer premises. The integration of data from a variety of customer meters communicating in different protocols to the collecting hubs and further communication and processing of the data at the substation are challenges to be addressed.

1.6.3   Types of data and signals in power system

In the monitoring part of the SCADA systems, the data acquired can be broadly classified into two categories: analog and digital. Pulse data also are acquired, as per the requirement, in case of a count accumulation function, like energy meter data.

1.6.3.1   Analog signals

Analog data involve all continuous, time-varying signals from the field, and are usually thought of in an electrical context; however, mechanical, pneumatic, hydraulic, and other systems may also convey analog signals. Examples are voltage, current, pressure, level, and temperature, to name a few. In power systems, the voltage transformers step down the voltages from kilovolt level to 110 V, and the voltage transducer converts the physical signals to milliampere current (normally 4 to 20 mA) range which is then used for further transmission. Current output is preferred for transducers due to the ease of transmission over long distances and because it is less prone to distortion by interferences. The output of a transducer that measures the power is shown in Figure 1.5 where the range is from 4 to 20 mA. The threshold value of 4 mA was chosen for two reasons. The first reason is that zero input corresponds to 4 mA, not zero amperes, which helps to identify a broken wire, which also will manifest as a zero output. The other reason is that the output curve of the transducer is linear along the 4 to 20 mA portion, as seen from the figure, which gives an accurate output.

Errors are introduced in the measurement due to the saturation of the current and voltage transformers, which creates a major problem in

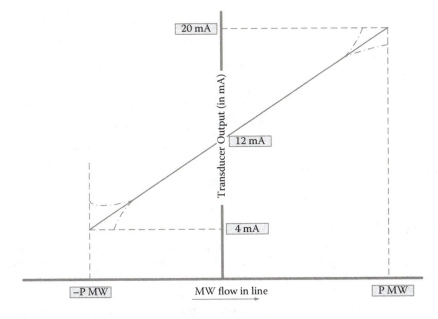

Figure 1.5 The transducer output curve at 4 to 20 mA.

magnitude and phase angle measurements. Errors can also occur due to the poor precision levels of the instrument transformers. The characteristics can deteriorate with time, temperature, and environmental factors.

1.6.3.2 Data acquisition systems

Data acquisition is the process of sampling real-world physical conditions and converting the resulting samples into digital numeric values that can be manipulated by a computer. Data acquisition and data acquisition systems (DASs) typically involve the conversion of analog waveforms into digital values for processing. The components of data acquisition systems include the following:

- Sensors/transducers that convert physical parameters to electrical signals (4 to 20 mA generally)
- Signal conditioning circuitry to convert sensor signals into a form that can be converted to digital values
- Analog-to-digital converters, which convert conditioned sensor signals to digital values

Figure 1.6 depicts a typical analog-to-digital conversion circuitry block diagram.

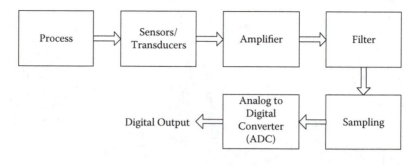

Figure 1.6 Analog to digital conversion.

1.6.3.3   Digital signals

A digital data signal is a discontinuous signal that changes from one state to another in discrete steps, usually represented in binary, or two levels, low and high. Digital signals include switch positions and isolator and circuit breaker positions in a power system.

Digital signals can be directly accessed by the automation system; however, for physical isolation, all digital signals come into the system via interposing relays. Interposing relays initiate action in a circuit in response to some change in conditions in that circuit or in some other circuit, as illustrated in Figure 1.7. Potential free contacts are used for bringing the data from the field, as can be seen in Figure 1.7. The coupling is electromagnetic from the circuit breaker contacts to the RTU, so that no

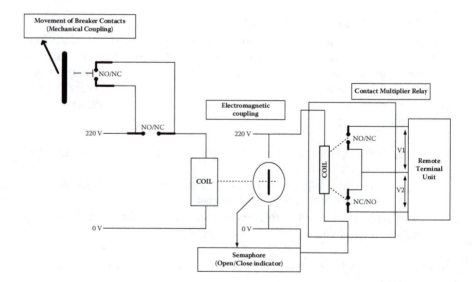

Figure 1.7 Status point data acquisition.

physical wiring from the field reaches the control equipment. Errors can be introduced here due to the rusting of the contacts or maloperation.

1.6.3.4   Pulse signals

Pulse data refer to the periodic information to be acquired from the field. Pulse data capture the duration between the changes in the value of a signal. This includes the energy data, rainfall, and so forth, and the outputs could be the stepper motor pulse signals.

1.7   Flow of data from the field to the SCADA control center

The flow of information in a SCADA system can be tracked by analyzing the flow of an analog signal from the field to the display screen of a dispatcher, as shown in Figure 1.8. As an example, the display of a bus bar voltage, say 220 kV, on the mimic screen of an operator is illustrated.

Starting from the substation bus bar in the field, the potential transformer connected to the bus converts the 220 kV into 110 V. This 110 V is converted into a 4 to 20 mA analog signal by a voltage transducer. As explained, this analog signal needs to be converted to a digital signal for onward transmission to the master station. The 4 to 20 mA analog signal is converted to a digital signal by the analog input (AI) module of the RTU. Further, this digital signal obtained is packaged into a data packet in the RTU, according to the communication protocol existing between the RTU and the master station. The data packets are then transmitted to the master station along the communication medium available. In the master station, the packets are received by the front-end processor/communication front end (FEP/CFE), decoded, and the data retrieved. The data are then scaled up to the 220 kV range and displayed at the appropriate bus bar in the mimic diagram of the operator console, completing the *monitoring* cycle.

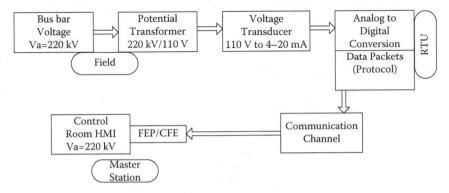

Figure 1.8 Data transfer from the bus bar to the control center HMI.

The same sequence could be retraced from the master station to the field, in the case of a control command issued by the operator, to be executed in the field.

Use of RTUs with hardwired I/O and serial communications, once predominant with all field equipment, has transitioned to data concentrators talking to IEDs with digital networked communications. Where substations used to have 100% of their points hardwired, new substations today have 5% or fewer of their points hardwired. Thus, the transition from the conventional RTU to the data concentrator can be seen.

1.8   Organization of the book

The book is organized into seven chapters, as given in Figure 1.9. Chapter 1 discusses the history of automation systems and how the SCADA control centers evolved. Chapter 2 elaborates the fundamental building blocks of SCADA systems including RTUs/IEDs, master stations, and human-machine interface in detail. Chapter 3 discusses the SCADA communication with emphasis on the protocols, media usage, and requirements. The rest of the chapters deal with the application of SCADA and associated technologies to the power system. Hence, Chapter 4 discusses substation automation, which is the SCADA application at the substation level, associated with any kind of substation, whether it is at the generation

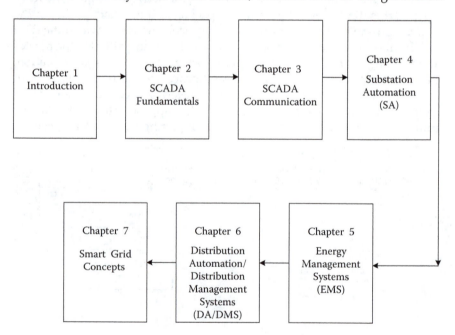

Figure 1.9 Organization of the book.

switch yard, transmission, or distribution level. Chapter 5 discusses the range of application functions associated with the basic SCADA systems, when applied to transmission systems, termed *energy management systems* (EMS). The SCADA and associated applications for distribution systems form the content of Chapter 6. Chapter 7, the concluding chapter, introduces the smart grid concept and the functionalities to be integrated and the challenges ahead.

1.9   Summary

This chapter is an introduction to SCADA systems, the history of power system automation, and the use of SCADA in the power sector. The advantages and application of SCADA in the power sector are discussed, and the types of data available at the power system and customer are touched upon for clarity.

Bibliography

1. J. D. McDonald, Substation automation, IED integration and availability of information, *IEEE Power & Energy Magazine*, vol. 1, no. 2, pp. 22–31, March/April 2003.
2. Mini S. Thomas, Pramod Kumar, and V. K. Chandna, Design, development and commissioning of a supervisory control and data acquisition (SCADA) laboratory for research and training, *IEEE Transactions on Power Systems*, vol. 20, pp. 1582–1588, August 2004.
3. IEEE Tutorial course on Fundamentals of Supervisory Systems, course 94 EH0392-1 PWR.
4. James Momoh, *Electric Power Distribution, Automation Protection and Control*, CRC Press, Boca Raton, FL, 2007.
5. Mini S. Thomas, D. P. Kothari, and Anupama Prakash, IED models for data generation in a transmission substation, in *Proceedings of the IEEE Conference: PEDES-2010*, December 2010, pp. 1–8. DOI: 10.1109/PEDES.2010.5712415.
6. James Northcote-Green and Robert Wilson, *Control and Automation of Electrical Power Distribution Systems*, CRC Press, Boca Raton, FL, 2006.
7. William T. Shaw, *Cybersecurity for SCADA Systems*, Pennwell, Tulsa, OK, 2006.
8. IEEE Tutorial Energy Control Center Design, course 77 TU0010-9-PWR.
9. Mini S. Thomas, Anupama Prakash, and D. P. Kothari, Design, development and commissioning of a substation automation laboratory to enhance learning, *IEEE Transactions on Education*, vol. 54, no. 2, pp. 286–293, May 2011.
10. Mini S. Thomas, Remote control, *IEEE Power & Energy Magazine*, vol. 8, no. 4, pp. 53–60, July/August 2010.
11. John D. McDonald, *Electric Power Substations Engineering*, 3rd ed., CRC Press, Boca Raton, FL, 2012.

chapter two

SCADA fundamentals

2.1 Introduction

Supervisory control and data acquisition (SCADA) systems are extensively used for monitoring and controlling geographically distributed processes in a variety of industries. However, many of the SCADA-related products are proprietary, and the knowledge of the components is acquired by the personnel on the job. Hence, students and new graduates find it difficult to understand the fundamentals of SCADA systems. An attempt has been made in this chapter to elaborate on the essential components of the SCADA systems which will help explain the functioning and the hierarchy, especially for power systems.

2.2 Open system: Need and advantages

SCADA systems are complex and require a variety of hardware and software seamlessly integrated into a system that can perform the monitoring and control operation of the large process involved. Communication among devices is key to successful SCADA implementation in modern power systems. Traditionally most vendors in the automation scenario established their own unique ("proprietary") way to communicate between devices. Getting two vendors' proprietary devices to communicate properly is a complex and expensive task. The possible solution to the problem is through two basic approaches:

1. Buy everything from one vendor.
2. Get vendors to agree on a standard communication interface.

The first proposition was widely used as earlier proprietary products were utilized for SCADA implementations and large turnkey projects were commissioned by a single vendor. This created a monopoly of products and processes, and it became increasingly difficult to maintain or expand the established SCADA systems.

The latter approach, to get all the vendors to agree on a standard communication interface, is the fundamental objective of the "open systems" movement. This led to the concept of nonproprietary, open systems, which created a level playing field for all the players in the automation industry. Interoperable systems are becoming popular due to the huge advantages they provide for manufacturers, vendors, and end users.

An open system is a computer system that embodies vendor-independent standards so that software may be applied on many different platforms and can interoperate with other applications on local and remote systems.

Open systems are thus an evolutionary means for a control system, based on the use of nonproprietary and standard software and hardware interfaces, that enables future upgrades to be available from multiple vendors at lowered cost and integrated with relative ease and low risk.

The advantages of open systems are manifold, evolving from the definition:

- Vendor-independent platforms for project implementation can be used, avoiding reliance on a single vendor.
- Interoperable products are used. Turnkey projects where one vendor supplies and implements the complete project are no longer required, as use of hardware and software from different vendors is possible.
- Standard software that could be used to program different hardware can be used.
- The *de jure* (by law) and *de facto* (in fact or actually) standards can be used.
- System and intelligent electronic devices (IEDs) from competing suppliers will have common elements that allow for interchange and the sharing of information.
- Open systems are upgradable and expandable.
- They have a longer expected system life.
- There are readily available third-party components.

As the following sections discuss the building blocks of SCADA systems, it is apparent that all the components discussed use open systems now and the SCADA implementations are exciting propositions with hardware and software acquired from multiple vendors as per the functional requirements of each system.

2.3 Building blocks of SCADA systems

The SCADA system has four components, the first being the remote terminal unit (RTU) or data concentrator, which is the link of the control system to the field, for acquiring the data from the field devices and passing on the control commands from the control station to the field devices. Modern-day SCADA systems are incomplete without the data concentrators and intelligent electronic devices (IEDs) which are replacing the conventional RTUs with their hardwired input and output (I/O) points. In this book, both RTUs and IEDs have been discussed in detail. Legacy systems with only RTUs, hybrid systems with RTUs and IEDs, and new systems with only IEDs have to be handled with ease by the SCADA system designer today. The second component is the communication system that carries the monitored data from the RTU to the control center and the control commands from the master station to the RTU or data concentrator to be conveyed to the field. The communication system is of great significance in SCADA generally and in power automation specifically, as the power system field is widely distributed over the landscape, and critical information that is time bound is to be communicated to the master station and control decisions to the field. The third component of the SCADA system is the master station where the operator monitors the system and makes control decisions to be conveyed to the field. The fourth component is the user interface (UI) also referred to as the human-machine interface (HMI) which is the interaction between the operator and the machine. Figure 2.1 gives a pictorial representation of the components of a SCADA system. All automation systems essentially have these four components, in varied proportions depending on the process requirements. Power system SCADA systems are focused on the master stations and HMI is of great significance, whereas process automation is focused on controllers, and master station and the HMI has less significance. The following sections will elaborate how the components of the SCADA system work cohesively to accomplish monitoring and control of the process to achieve optimum performance of the system.

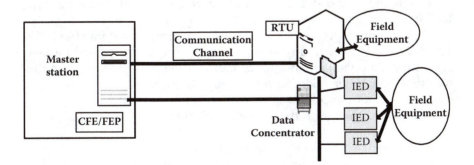

Figure 2.1 Components of SCADA systems.

2.4 Remote terminal unit (RTU) [1–7,18–19,24]

The RTU is the eyes, ears, and hands of the SCADA system. In older days, RTU was a slave of the master station, but now RTUs are equipped with internal computational and optimization facilities. RTU collects data from the field devices, processes the data, and sends the data to the master station through the communication system to assist the monitoring of the power system as "eyes" and "ears" of the master station. At the same time, the RTU receives control commands from the master station and transmits these commands to the field devices, thus justifying the comparison to the "hands" of the master station. Figure 2.1 shows the location of the RTU and the communication front end/front-end processor (CFE/FEP) of the master station.

2.4.1 Evolution of RTUs

From 1900 to the early 1920s, varieties of remote control systems were developed by engineers for remotely supervising processes. The systems could only monitor the process and no control was possible. In 1921 a system designed by John B. Harlow could automatically detect a change of status at a remote station and could report the change to the control center.

In 1923 the remote control system developed by John J. Bellamy and Rodney G. Richardson employed an equivalent of our modern "check before-operate" technique. It ensured the validity of a selected control point before the actual control was initiated. In 1927 the first logging system, designed by Harry E. Hersey, monitored information from a remote location and printed status change with reported time and date.

Supervisory systems evolved from electromechanical to using solid-state components, electronic sensors, and analog-to-digital converters. With the advent of microprocessors, RTU manufacturers merely upgraded their technology and did not look at alternate ways of performing the RTU function.

In 1980s, microprocessor-based logic was incorporated into the RTUs. This increased the flexibility of supervisory systems and brought in new capabilities in operation and performance. The development in communications and faster microprocessor chips brought down the costs and improved performance.

The new systems had the following advantages:

1. Modular system development capability
2. Largely preprogrammed user interface system that is easy to adapt to the individual process
3. Preprogrammed menu-driven software (final programming using a few buttons on keyboard)

4. Wide selection of control algorithms with preprogrammed menu
5. Data highway with transmission and communication capabilities between separate units—wideband, redundancy
6. Relatively easy communication with the control room for supervisory control
7. Extensive diagnostic scheme and devices for easy maintenance and replacement of circuit board (card level)
8. Redundancy at any level to improve the reliability
9. Industry standard communication protocols (IEEE 1815 or DNP3, IEC 60870-5-101 and 103)

2.4.2 Components of RTU

RTU has the following major components to accomplish the tasks of monitoring and controlling the field devices:

1. *Communication Subsystem*: Communication subsystem is the interface between the SCADA communication network and the RTU internal logic. This subsystem receives messages from the master, interprets the messages, initiates actions within the RTU which in turn initiates some action in the field. RTU also sends an appropriate message to the master station on the completion of the task. It also collects data from the field, and processes and conveys relevant data to the master station. RTU may report to a single master or multiple masters.
2. *Logic Subsystem*: The logic subsystem consists of the main processor and database and handles all major processing—time keeping, and control sensing. The logic subsystem also handles the analog-to-digital conversions and computational optimization, in most of the cases.
3. *Termination Subsystem*: The termination subsystem provides the interface between RTU and external equipment such as the communication lines, primary source, and substation devices. RTU logic needs to be protected from the harsh environment of the substation.
4. *Power Supply Subsystem*: The power supply converts primary power, usually from the substation battery, to the supply requirements of the other RTU subsystems.
5. *Test/HMI Subsystem*: This subsystem covers a variety of components, built-in hardware/firmware tests, and visual indicators, within the RTU, and built-in or portable test/maintenance panels or displays.

Figure 2.2 shows the components of the RTU, and the following sections will provide details of each of the RTU components. Figure 2.3 presents a typical RTU in a substation.

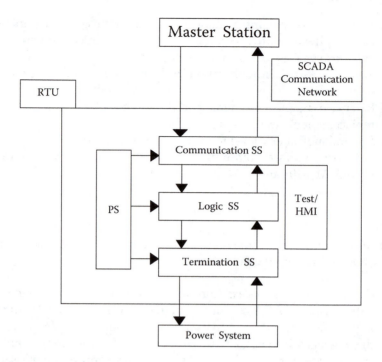

Figure 2.2 Components of RTU.

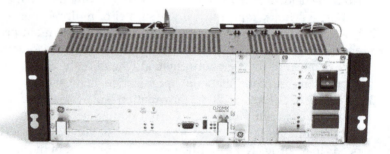

Figure 2.3 Typical RTU in a substation. (Courtesy of GE.)

2.4.3   Communication subsystem

The communication subsystem is the interface between the SCADA communication network and the RTU internal logic. Messages from the master station are received and interpreted by the communication subsystem, and the required action is initiated within the RTU. The RTU then initiates the requisite control action in the field, on the completion of which an appropriate message is transmitted to the master station. The communication subsystem receives data from the field, processes the data, bundles the relevant data in the appropriate protocol, and conveys the data to the master station, via the SCADA communication network. Hence, it is evident that the communication subsystem of the RTU is responsible for interpreting the messages from the master station, as well as formatting the messages to be transmitted to the master, including the message security. The RTU communication subsystem handles the following functions

2.4.3.1 Communication protocols

A large variety of communication protocols exist in the power system, and the RTU communication system is designed to format and interpret the data in the required protocol. Details of the communication protocol structure and the protocols used in the power system are discussed in Chapter 3. SCADA communication protocols generally "report by exception" or give information on the points that have changed since the last scan, to reduce the communication system load. For analog points, this means changing beyond their deadband between scans.

2.4.3.2 Message security

The data handled by the SCADA system are critical, and any corruption in the data can lead to serious consequences. Parity check is the simplest method, where a single bit is added to the message so that the sum is always odd. Cyclic redundancy check (CRC) is another error-checking mechanism used, which is more reliable. Here, each block of data is divided by a 16-degree polynomial; the remainder of the division is added to the end of the message block. The message will have a fixed length preamble of overhead characters, depending on the protocol used, the station address, the function code, and other details. CRC code is calculated separately for the preamble and the data block.

2.4.3.3 Multi-port communication

Modern RTUs have to communicate to the higher SCADA hierarchy to more than one master station, and at the same time, communicate with peer RTUs and IEDs in a variety of protocols. The communication subsystem should be designed to handle this capability.

2.4.4 Logic subsystem

The logic subsystem is the central processing and control unit of the RTU. Modern-day RTUs perform a number of advanced functions to off-load the master station in addition to the two primary functions: data collection and processing and control point selection and execution.

The primary functions of the RTU are time keeping and data acquisition and processing, as shown in Figure 2.4.

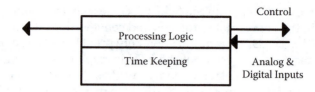

Figure 2.4 RTU logic subsystem (basic).

2.4.4.1   Time keeping

Sequence-of-events (SOE) logging or time tagging of events is of great significance in power systems, and the logic subsystem handles this task in the RTU. The RTU also has to perform many functions on a time basis. The RTU supports time synchronization in addition to time keeping. Time synchronization of the RTU and the master station takes place through the GPS receivers which ensure perfect synchronism (1 ms resolution). Some of the RTUs are time synchronized from the master station (±6 to 8 ms resolution, at least). For effective sequence-of-events logging, the RTU must be able to measure time to within 1 ms. This degree of precision requires both an accurate clock and an interrupt-driven processor.

2.4.4.2   Data acquisition and processing

SCADA data come in analog and digital forms. The logic subsystem data acquisition processing collects and reports both types of data. The analog values are acquired from transducers connected to the field devices, an example being the current and voltage values from the transmission lines or transformers. The earlier generation of RTUs had the analog-to-digital conversion module as part of the RTU, requiring hardwires to be brought in from the field to the RTU. With the advances in analog-to-digital conversion techniques and communication networks, field devices are becoming "intelligent" and can supply digital data directly to a LAN which in turn can be acquired by the RTU. Presently, IEDs are deployed extensively in the field for SCADA applications, and transmit all data in digital form. Data acquisition was discussed in Section 1.6.3.

2.4.4.3   Digital data acquisition

Digital data is a status or contact input which has two states and is generally received as an actual contact/switch closure or a voltage/no-voltage signal. This signal indicates the current status of a system, such as on/off or open/close.

Digital data acquisition is done in four ways:

1. Current status
2. Current status with memory detect—number of contact changes since last report

3. Sequence of events (SOE)—with time tag
4. Accumulator value—a count of the number of contact closures over a period of time (generally used for energy meter pulse generators)

Digital data acquisition is generally done in two ways, high-speed scanning of all input points, whereas some RTUs scan the analog input points and use microprocessor interrupt for status changes. Only RTUs with microprocessor interrupt are suitable for sequence-of-events (SOE) logging and can time-tag events to 1 ms accuracy.

The SOE logs are used by system planning and/or system protection engineers to verify that the protection system worked as designed.

2.4.4.4   Analog data acquisition

The analog signals, generally a voltage or current that changes over a period of time and also within a certain range, are generally converted to a 4 to 20 mA signal by the appropriate transducers. Some utilities also use –1 to +1 mA. The analog-to-digital converter circuit converts these signals into binary values for further transmission or analysis by the RTU. The analog signals should be free of noise and electromagnetic interference. The 4 to 20 mA current loop signals are generally immune from electrical noise sources and are the most preferred standard input to the A/D converters. The 4 mA threshold is given to account for a break in the circuit, which will show a zero, and the minimum measured value will show 4 mA. The logic subsystem of the modern RTU is also capable of handling many more functions such as filtering, linearizing, report by exception, and limit checking/alarming, to off-load the communication channel and the master station. The multiplexing of inputs is done to utilize the capability of the A/D converters. With the microprocessor-based systems in place, selected inputs can be scanned more frequently and facilitates the fetching of critical analog inputs more frequently.

2.4.4.5   Analog outputs

Analog outputs are used to vary the operating points of process variables, such as the change of the level, variable speed motors and drives. This will generally be a constant milliamp signal proportional to the digital quantity specified in a command from the master station. Analog outputs used to drive strip chart recorders in the control center are rarely used by the power system automation industry.

2.4.4.6   Digital (contact) output

Contact outputs are the control command issued by the RTU for opening or closing of any kind of switch, whether a circuit breaker, isolator, or simple switch. This is generally achieved by an electromechanical relay (interposing relay) operation, initiated by the RTU, to operate equipment. The process industry and power systems use this kind of output extensively. Contact outputs are latched and remain in that position until another specific command is given. Momentary outputs are turned on and off after a specific defined time by a single command. In modern RTUs, the on-time is user defined and is variable, whereas in the earlier RTUs, the time was preset.

In power systems, contact outputs support separate trip and close relays, and there is a select-before-operate (SBO) security provision, and circuitry to ensure that one, and only one, control relay in the RTU is operated. The SBO feature provides for an end-to-end system check whereby the master station sends a point selection message and receives a checkback message before sending an operate message and receiving a verification message. The control logic completely resets upon completion of an operation or when a defect is detected.

2.4.4.7   Pulse inputs

Pulse inputs yield a numeric value, like the analog inputs; however, they are considered as a special class and have two classifications—pulses that are continuously counted and pulses that are counted over a specific time interval. Each pulse will represent a specific quantity of an input, such as 1 mm of rainfall, a kilowatt of power, and so forth. Thus by counting the pulses, the total quantity over a period of time can be arrived at, the rainfall during a day, the power consumed for specific hours, and so on. Pulse inputs are used in power system SCADA for energy accounting, with inputs from the power meters. Modern RTUs require no special hardware, as the pulse inputs can be sensed by the status input hardware.

2.4.4.8   Pulse outputs

Pulse outputs are a special class of contact outputs where the output changes between on and off for a specific number of times. These can be considered as a special class of digital outputs. They were mainly used by process industries and are now rarely used, but this section is included for the sake of completeness of the components.

2.4.5   Termination subsystem

The termination subsystem is the interface between the RTU, which is an electronic device, and the physical world, which is generally hazardous for the RTU. The main function of the termination subsystem is to protect the RTU from the hostile field environment. The substation environment is hostile due to many factors such as surges, lightning, over voltage and reverse voltage, electrostatic discharge (ESD), and electromagnetic interference (EMI). In the case of process industry, the hazardous environment will include temperature, humidity, and fumes. The actual provisions of isolation between the RTU logic subsystem and the field will depend upon individual manufacturer; however, the bottom line is the RTU will have to be protected from the hazardous environment.

2.4.5.1   Digital terminations

The digital inputs to the RTU are from the various forms of switches in the field and originate from the electromechanical contacts in switchgear and metering devices.

Generally the contact sensing is done by an interposing relay powered by a battery, which provides isolation from the field. Optical isolators are also prevalent, which provide complete isolation. If the contact input comes from a metering device, then the firmware contains change detection and pulse accumulation logic.

2.4.5.2   Analog terminations

Analog inputs are from the transducers, sensors, transmitters, thermocouples, and resistance devices, which themselves provide electrical isolation. The 4 to 20 mA signals from these devices reach the analog-to-digital conversion unit through fuses and are grounded at the RTU.

2.4.6   Testing and human-machine interface (HMI) subsystem

RTUs located at remote locations are generally unmanned and may not have a display system or HMI associated with them. The panel of the RTU will have a number of LEDs which indicate the status of the various cards and functionalities of the RTU, which give the personnel an idea about the status of the RTU. Figure 2.5 shows a typical RTU with indicators.

The RTU will have its own built-in routines that can test the hardware and software and give indications on the panel. The test results and related information will be passed on to the master. Continuous monitoring of the firmware and software of the RTU is done so that faults and problems can be identified and rectified instantly. Card-level diagnostics software can be run at the master station level to identify the faulty cards and appropriate corrective action can be initiated. Plug-in test sets that can simulate a master station and test the RTUs are also used by technicians for better diagnosis of RTU-related problems.

With the availability of low-cost LED and LCD displays, RTUs can be fitted with such display panels that will give the values measured by the RTU to convey information to the personnel present in the plant floor or substation if necessary.

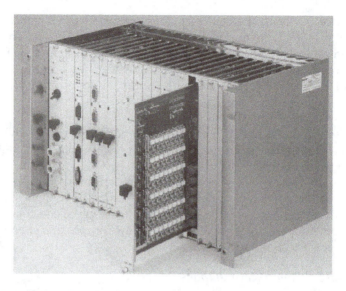

Figure 2.5 RTU with draw-out cards for communication, analog and digital inputs and outputs, and logic and power supply units. (Courtesy of PGCIL Ltd, India.)

2.4.7   Power supplies

The RTU will have a separate power supply unit, which is powered from a suitable DC source. The most common voltage levels in use are 24 VDC, 48 VDC, and 125 VDC. Sometimes even 250 VDC may be used in systems. RTUs in the transmission and distribution system are located in the substation and are powered from the substation battery. These batteries are floating so that a single fault on either side of the battery to ground will not cause malfunction, equipment damage, or danger to human beings. Many premises will have two voltage levels, say 24 VDC and 48 VDC, and the RTU supply can be easily switched from one to the other, making the system more reliable.

2.4.8   Advanced RTU functionalities

With the advent of microprocessor technology and with the integrated circuit–based devices becoming cheaper by the day, the RTUs also gained in functionality and versatility. The CPUs became faster, with more memory and advanced computations possible. The RTU feature evolution also varied for different industries as per the market segment. In the power industry, the major advancements have been in the following aspects, as shown in Figure 2.6.

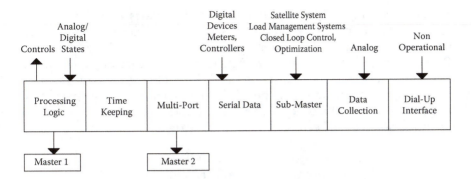

Figure 2.6 Advanced RTU functionality of the logic subsystem.

2.4.8.1 Multi-port and multi-protocol operation

When two master stations required data from the same substation, the inclination was to install two RTUs, each reporting to the respective master stations, probably the same parameters. With logic systems migrating to microprocessor-based systems, the memory and processing power increased and the same RTU could report to more than one master station. Sometimes, RTUs at the same hierarchical level will have to communicate with each other to implement certain control functions, and this also required multi-port communication capability. The cost and complexity of the firmware in the logic and communication subsystems will increase in this case; however, the total cost of installing and operating the system will be greatly reduced.

The RTU logic subsystem will have a general database to store all information from the I/O units, and secondary databases are created to cater to each master station for the points that are monitored and controlled by that particular station. The protocol used for communication between separate master stations also could be different; hence, the communication subsystem of the RTU which interprets and formats messages should be able to handle multiple protocols.

As an example, the RTU at a substation in a transmission SCADA hierarchy will report to the local master station as well as the regional control center. It may also coordinate with another RTU at the local level, which may be for implementing protection functions.

2.4.8.2 Digital interface to other electronic devices

The electrical utilities presently use many intelligent electronic devices, with data acquisition and processing capability for control, protection, and metering applications. The data from these devices can be integrated to the RTU for processing and onward transmission higher in the hierarchy.

Each of these IEDs will have different electrical interfaces and communication protocols; hence, the RTU has to be equipped to handle these digitized data. Generally serial interfaces are used which include RS 232and RS 485 for such communication, and there is a limit regarding the number of serial ports that are supported in the RTU and the number of IEDs that can be supported. In addition, the RTU is able to support only operational data, whereas the data concentrator can support both the operational and nonoperational data. See Section 4.4 for more details.

2.4.8.3 *Closed-loop control, computation,* *and optimization at the RTU level*

Modern RTUs can handle closed-loop control, complex computations, and optimization. These are used extensively to off-load the master station and to have distributed control. For closed-loop control, the set point value is accessed from the master, the measured value is compared, and the appropriate action is initiated by the RTU to maintain the measured value equal to the set value. The examples are the on-load tap changers with power transformers and voltage regulators, where the tap position is changed to maintain the voltage at the specific set point value.

Complex computations include using measured values of many parameters to compute a value, which will be the set point and the resulting control action. For controlling the capacitor banks along a distribution feeder for power factor correction, voltage improvement, and loss reduction, many line parameters like voltage, power factor, and reactive power flow are measured and computed before the switched capacitor banks are controlled. An optimization algorithm can be used as the next level to optimize the measured or calculated quantity.

It is clear that these actions reduce the burden on the master station, and the RTUs are equipped for such complex functions by distributing the intelligence.

2.4.8.4 *Interface to application functions*

At the substation level, many utilities have installed application algorithms for optimization of the operations, like load management and now demand response systems. The operator can initiate the load reduction and other activities using the same system, and the RTU has to establish and interface with such application programs.

2.4.8.5 *Advanced data processing*

The RTUs collect and transmit a large number of data points, and the scan time of each point is in the range of 2 to 10 s (e.g., 2 s for digital values and 10 s for analog inputs) which gives an idea about the amount of data received at the master station or higher hierarchy. The operators are flooded with data which leads to requests to implement advanced processing features into the RTU. One of these requests is to consider analysis of status point changes to report only the high-level action. For example, when a circuit breaker operates, numerous analog point alarms can be generated (low voltage and low current for all three phases). The important message to give the system operator is the fact the circuit breaker operated (primary alarm), but present systems also provide the analog point alarms (secondary alarms), too. The desire is to log all the analog point changes in the master station for later engineering analysis but report only the breaker operation to the system operator. Intelligent alarm processing is advanced data processing that can be installed in the RTU (see Section 4.9.3.2).

2.4.8.6   Other functions

Time tagging of analog and digital values for sequence of events recording is implemented in RTUs. Other functions that can be implemented in the RTU include distribution automation, volt-ampere reactive (VAR) control and fault detection, isolation, and service restoration, as explained in later chapters.

Thus, it is evident that modern-day RTUs are power houses that, in addition to acquiring the data from the field and executing the control actions, are capable of performing a variety of other functions and are integral components of the SCADA system.

2.5   Intelligent electronic devices (IEDs)

The industry standard definition of an IED is "Any device incorporating one or more processors with the capability to receive or send data/control from or to an external source (e.g., electronic multifunction meters, digital relays, and controllers)."

IEDs have been deployed extensively in power automation systems recently, and the shift from RTUs to IEDs is evident due to the integration and interoperability features of the IEDs. It is necessary at this point to discuss the IED functionality in detail to present a holistic view of automation in power systems.

2.5.1   Evolution of IEDs

IEDs were introduced in the early 1980s with microprocessor-based control features. The deployment of IEDs is revolutionizing the protection, substation and distribution automation, and data capture and analysis functions of an electric utility. The protection relay migrated from single-function conventional electromechanical types to multi-function microprocessor-based relays and started incorporating different protection functions into the same relay, rather than using individual relays for each application. Considerable savings were achieved in relay panel and switchgear costs by the adoption of the multi-function microprocessor-based relays.

However, IED revolution started when other functionalities like accurate voltage and current phasor measurement, waveform capture, and metering were being incorporated into the relays.

The growth in communication infrastructure, standardization of protocols, and interoperability were major factors that led to the IED explosion. IEDs are now the eyes, ears, and hands of the automation systems in a power utility. IED packed with full control and monitoring capabilities and with analysis of fault report data can manage substations without human intervention. The wrong tripping of circuits can be avoided by utilizing the IEDs' capabilities to the full extent.

With highly integrated IEDs, utilities and industrial plants have a huge potential for cost savings. These savings can be summarized in the following categories:

1. Lower installation and panel assembly cost
2. Shorter commissioning and maintenance times
3. Shorter system recovery time after a disturbance
4. Less revenue loss due to wrong settings and IED malfunction
5. Higher system reliability due to automation, integration, and adaptive settings
6. Better utilization of installed capacity
7. Better justification of new investments
8. Smaller control houses

Integration of IEDs and proper analysis of fault data will lead to very short system restoration times after a blackout, and revenue losses of utilities will be minimized.

2.5.2   IED functional block diagram

Figure 2.7 depicts the structural block diagram of a typical intelligent electronic device. The modern IED architecture ensures that the device is multipurpose, modular in nature, flexible and adaptable, and has robust communication capabilities. Communication capabilities include multiple selectable protocols, multi-drop facilities with multiple ports, and rapid response for real-time data. IEDs also have tremendous data-processing capability for a variety of functions, for various applications like protection and metering. IEDs have event recording capability that can be very

External Communication	Data Processing	Input/Output Measurement
Selectable Protocol	Protection*	Discrete Inputs
Selectable Protocol	Metering	Analog Inputs*
Rapid Response	Event Recording	Discrete Outputs*
Real-Time Data*	Fault Recording	Analog Outputs*
Multiple Ports	Application Logic	Selectable Ratings

*Old Relay

Figure 2.7 Modern IED with the functional blocks.

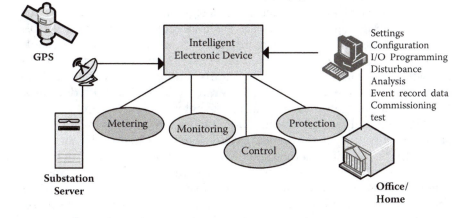

Figure 2.8 (See color insert.) Functional view of modern IED.

useful for post-event analysis, for fault waveform recording, and for power quality measurements. This eliminates additional digital fault recorders and power quality monitors. IEDs can also accept and send out analog and digital signals with selectable ratings, thus making the IEDs versatile.

As far as IED deployment in the field is concerned, Figure 2.8 gives a complete picture of the IED integration with the devices connected and the functionalities handled by the IEDs.

The IED brings a relay panel with many single-function electromechanical relays, control switches, extensive wiring, and much more into a single box. In addition the IED handles additional features like self and external circuit monitoring, real-time synchronization of the event monitoring, local and substation data access, programmable logic controller functionality, and an entire range of software tools for commissioning, testing, event reporting, and fault analysis. Typical relay IEDs are shown in Figure 2.9.

The following sections will elaborate the IED building blocks in detail.

2.5.3   Hardware and software architecture of the IED

The architecture of an IED should ensure the ease of use of the device in regard to programming, commissioning, and maintenance. The hardware should be designed with the future adaptability requirement in mind, whereas the software structure should ensure the independent protection, control, metering, and communication functions.

IED hardware design utilizes draw out–type cards which is a great advantage, as the replacement can be done easily without disconnecting the terminal wires and removing the IED from the panel.

The IED software architecture is designed in such a way that the commissioning engineer can easily evaluate and program the available functions independently. The required function can be selected, while the other functions are deactivated and will not be visible to the personnel, which helps to save time while commissioning. Each selected function is an independent embedded unit generally with the IED with dedicated logical inputs and outputs, setting, and event reporting features

Figure 2.10 illustrates the functional blocks in an IED which demonstrate the versatility of the device. In addition to the analog, digital inputs and outputs, the IED has the capability of waveform capture and disturbance analysis capability. Metering and demand values recording are other features, in addition to programmable logic capability of the IEDs that eliminates an additional PLC usage. Self and external circuit monitoring make the device extremely reliable and reduce downtime.

2.5.4   IED communication subsystem

IED communication is of utmost importance; hence, the device provides flexibility and at the same time major benefits to the utility. The IED should support different protocols for multi-port communication and different media and should have flexible and open communication architecture. HMI interface, remote access port and direct communication to other IEDs for protection purposes are musts for modern IEDs.

As discussed earlier, open protocols are the norm today, and the IEDs have plug-and-play communication modules that can support a variety of protocols. The advantage of these modules is that they can be replaced in

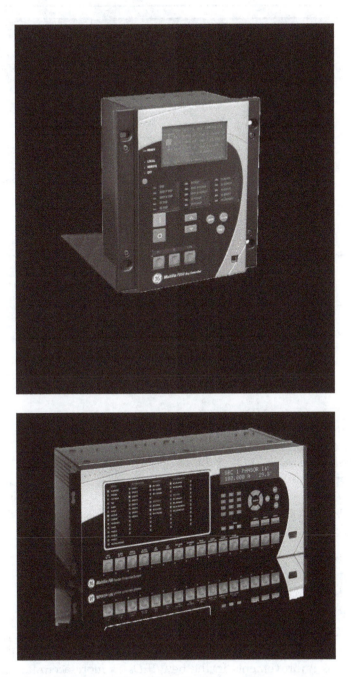

Figure 2.9 (See color insert.) Relay IEDs. (Courtesy of GE.)

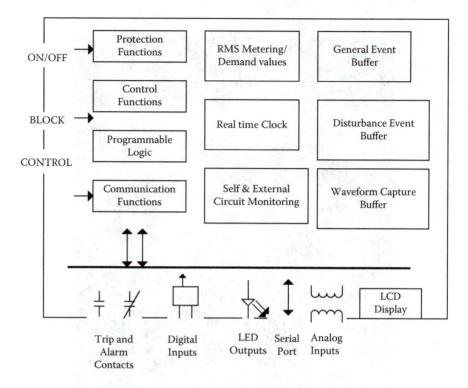

Figure 2.10 Structure of an IED [12].

the field in case of a change in communication requirement, and the IEDs can be integrated to a separate system at the higher hierarchy with ease. IEDs are capable of multi-port communication and can communicate with substations and other IEDs at the same time through a modem to office/home/service station.

The IEDs use the communication port and optical port for fiber optic communication or electrical port (RS-232 or RS-485) and will also have a service port for remote access via a modem.

Figure 2.11 shows the relay IEDs communicating to a computer/server at the substation.

2.5.5   IED advanced functionalities [11–15,17]

2.5.5.1   Protection function including phasor estimation
The protection function is the primary function of a relay IED, as IEDs are primarily the improvement on the microprocessor-based relays. There are tremendous improvements in the new IEDs as more accurate measurement principles and less auxiliary equipment are required. Auxiliary CTs can be eliminated in a transformer differential relay, as the new relay has

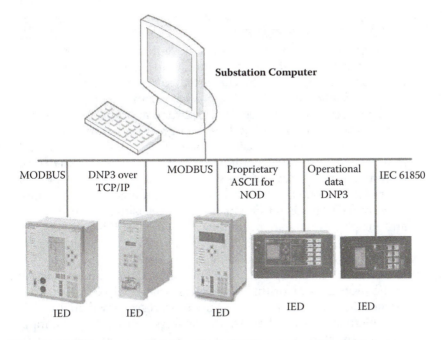

Figure 2.11 IED communication options (NOD, nonoperational data).

a CT mismatch correction function, as the original primary transformer currents are available for further analysis in a modern relay IED. Similarly, with appropriate techniques, and numerical comparison algorithms, the CT mismatch, inrush, and CT saturation problems can be solved without external devices.

McLaren was the first to propose the concept of open system relaying, where different relay functions can be obtained from the same hardware just by modifying microprocessor programming. Modern relay technology has shown recently a tendency toward this direction. The generalized numerical relay concept, which is directly derived from open system relaying, consists of a minimum set of hardware modules and functions of modern digital and numerical relays. With the generalized numerical relay and with the amount of information commonly available, it is possible to recreate the majority of modern digital and numerical relay equipment. The following data-processing modules constitute the generalized numerical relay:

1. *Isolation and analog signal scaling*: Current and voltage waveforms from instrument transformers are acquired and scaled down to convenient voltage levels for use in the digital and numerical relays.
2. *Analog anti-aliasing filtering*: Low-pass filters are used to avoid the phenomenon of aliasing in which the high-frequency components of the inputs appear to be parts of the fundamental frequency components.

3. *Analog-to-digital conversion*: Because digital processors can process numerical or logical data only, the waveforms of inputs must be sampled at discrete times. To achieve this, each analog signal is passed through a sample- and-hold module, and conveyed, one at a time, to an analog-to-digital converter (ADC) by a multiplexer.

4. *Phasor estimation algorithm*: A software algorithm implemented in a microprocessor estimates the amplitude and phase of the waveforms provided to the relay. This is of great importance in modern monitoring systems, as phasor measurements are becoming an integral part of the system monitoring. This feature of the IED is used to compute the phasor of the voltage or current with respect to a reference phasor. (The IEDs are time synchronized from a common GPS source.) This is termed as a phasor measurement unit (PMU), and the phasor data are accumulated by a phasor data concentrator (PDC) at an appropriate location, mostly in the control center. The phasor concept is explained in detail in Chapter 5.

5. *Relay algorithm and trip logic*: The equations and parameters specific to the protection algorithm and the associated trip logic are implemented in the software of the microprocessor used in the relay. The microprocessor calculates the phasors representing the inputs, acquires the status of the switches, performs protective relay calculations, and finally provides outputs for controlling the circuit breakers. The processor may also support communications, self-testing, target display, time clocks, and other tasks.

2.5.5.2 *Programmable logic and breaker control*

A modern relay IED eliminates the use of external programmable logic controllers (PLCs) as the IED can handle logical inputs and outputs of the protection functions, which can be connected to flip flops and/or gates of the IED directly.

Figure 2.12 presents an example of programmable logic tools. In Figure 2.13, an example of a PC-configuration tool for user programmable logic is shown.

2.5.5.3 *Metering and power quality analysis*

Metering capabilities of the IEDs became acceptable to the power utilities quickly, and major cost saving was achieved by combining the non-revenue metering function into the IEDs. It may be noted that the primary CTs and PTs for protection purposes may not be accurate enough for normal current measurement for revenue metering. The normal metering functions include measuring the voltage and current root mean square (RMS) values and the real and reactive power.

In addition to these basic functions, metering also includes the values for commissioning and testing, and this feature reduces the commission

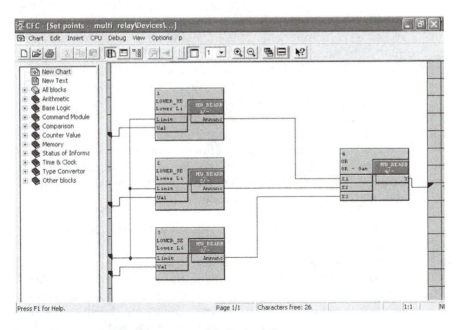

Figure 2.12 Example of programmable-logic tools.

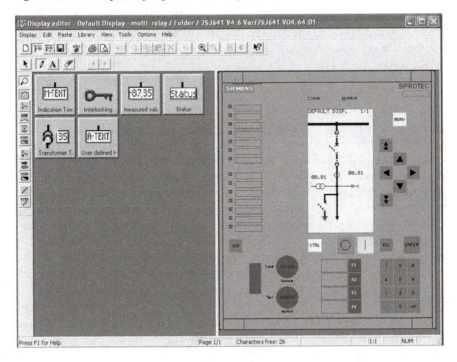

Figure 2.13 Programming of the graphic LCD display for breaker operation logic.

and testing times on the site. The metered values are the positive, negative, and zero sequence components of voltage and current phase shifts and the normal RMS values. The phase mismatch, differential, and restraint values can be computed easily to hasten the commissioning process.

Load profiling is another metering activity that can be achieved using IEDs where the power factor profile, ampere demand, long-term RMS voltage value, and so on, can be monitored and can be used for load profiling for long-term expansion planning.

Using the PLC and metering functions, many system requirements can be met without additional effort, like capacitor bank control by the reactive power data monitoring and control algorithm implemented using a PLC and programmable output contacts.

2.5.5.4   Self-monitoring and external circuit monitoring

IEDs have card-level diagnostic capabilities for internal problems, with a self-monitoring software, which can detect up to 98% of the problems, such as hardware failure, memory failure, and power supply problems. Modern-day IEDs, in addition to the internal monitoring, have capabilities for interface monitoring and external circuit monitoring. Interface monitoring includes checking the inputs to the IEDs and can be verified by simple methods. For example, the input currents to the relay from the three phases should add up to three times the neutral current if any. If there is any deviation, the analog channel of any of the currents could be faulty. The relay can block the false tripping. External circuit monitoring will include monitoring of the circuit breaker coil for any interruption in the trip-close path and can also indicate an instrument transformer failure.

2.5.5.5   Event reporting and fault diagnosis

Relay IEDs eliminate the digital fault recorders because waveform recording during a fault can be performed by the IEDs, whereas the electromechanical relays did not have such capability. Event reporting can be easily done by relay IEDs eliminating sequence of events (SOE) recorders. The relay IEDs save the captured data in nonvolatile memory and disturbance event reports (pick up, trip, and auto-reclose), and general event reports like changes of settings have to be saved and managed separately. Time stamping of all events is done by the IEDs, and GPS synchronization for this purpose and a battery backup for the real-time clock are essential. The events, once time tagged correctly, can be reported in the correct sequence in which they occurred, eliminating further sequencing at the control room. It is hence easy to perform fault diagnosis after a fault, as the values will be saved in the IED and can be retrieved later, even in case of a blackout.

2.5.6 Tools for settings, commissioning, and testing

User-friendly software tools are key to better planning, programming, commissioning, and testing of an IED. The elaborate avenues and application functionalities of an IED can be handled only by an intuitive, easy-to-use PC program.

IEDs come with menu-driven intuitive, easy-to-use, flexible software programming tools for settings and configurations during commissioning. These user-friendly programs come with factory presettings that simplify the job of the commissioning personnel.

2.5.7 Programmable LCD display

The programmable LCD display is a great tool in the new generation of IEDs. This is used for graphical information as well displaying text and can be switched between graphic and text modes. Figure 2.9 shows the display of relay IEDs. The topology of the bus and breaker including isolators, disconnecting switches, and many more configurations can be programmed using the software tools in the graphic mode. Text mode of the LCD display is used for settings and detailed display of the metering values in primary or secondary units. The LCD display can be switched between text and graphics modes.

2.5.8 Typical IEDs

IEDs, as discussed, are devices that can be connected to a LAN and communicate with other devices over the LAN and have processing capabilities. A large number of IEDs are available currently, relay IEDs being the most commonly used for automation purposes. However, the smart meter used for home automation is an IED and so is a programmable logic controller (PLC) used for automation. Digital fault recorders (DFRs) and remote terminal units (RTUs) are IEDs with digital data transmission and reception capability with computational facility built in.

Thus IEDs have become the basic building blocks for automation of power systems. However, it is not easy to replace all the existing RTU and related equipment with IEDs, and different approaches are used to integrate the legacy systems with the new systems. Before discussing the building of different kinds of SCADA systems, it is imperative to touch upon the data concentrators and merging units which are used along with RTUs and IEDs for data communication in the SCADA systems.

2.6 Data concentrators and merging units

RTUs were discussed in detail in Section 2.4; however, with the advent of IEDs, the way substations acquire data has evolved. RTUs get inputs from hardwires coming from the field and the analog to digital conversion of the values happen in the RTU. RTUs communicate to the higher hierarchy via any physical medium through the communication subsystem of the RTU.

2.6.1 RTUs, IEDs, and data concentrator

Data concentrators collect data from IEDs and other inputs from the field in a substation and can provide the complete or partial information to the higher hierarchy. The hardwires coming from the field carrying analog values and status points which terminate at the IED are processed to all digital values, as discussed earlier. Figure 2.14 shows the traditional RTUs and the modern IEDs with data concentrators implemented in a substation. The required information to be transmitted to the higher hierarchy is sent by the IEDs to the data concentrator using a communication protocol. The data concentrator communicates on a LAN as shown in Figure 2.14.

2.6.2 Merging units and IEDs

Merging units take the local area network to another level called the *process bus*, right into the field. The hardwired data from the field is brought into the merging unit, which is converted to all digital values by the merging units. The IEDs receive this data through the process bus LAN using a protocol, as shown in Figure 2.14. More details of the merging units and process bus are given in Sections 4.3.4 and 4.6.

2.7 SCADA communication systems

The communication system plays a vital role in the SCADA implementation, especially with many time critical applications in power systems. The communication media and the protocols used in the power system require special treatment and hence are detailed separately in Chapter 3.

2.8 Master station [5,6,8,24]

The SCADA master stations range from small control rooms in a substation to large transmission SCADA master stations manning the power flow of a whole country. The master station is a collection of computers, servers, peripherals, and I/O systems that help the operator to monitor the state of the field and initiate control actions at the appropriate moment.

The master station components can be classified into hardware and software components.

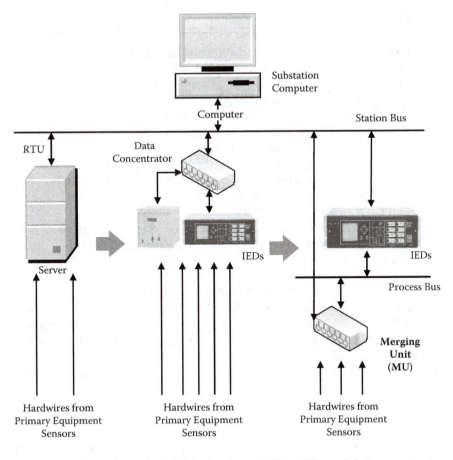

Figure 2.14 (See color insert.) Migration from RTU to IEDs and data concentrator to merging units and IEDs.

2.8.1   Master station software components

The master station software components can be broadly classified into two categories: the basic SCADA functions and advanced application functions pertaining to the specific SCADA implementation, like generation SCADA, transmission SCADA, or distribution SCADA applications.

2.8.1.1   Basic SCADA software

The basic SCADA software performs the basic functionalities of a SCADA system, and is common to all SCADA applications. Some of the major functions performed by the basic SCADA software are as follows:

1. *Data Acquisition and Control*: This includes the basic SCADA functions of data acquisition and control. This software has the basic modules for engineering and commissioning of projects.
2. *Database*: The power system SCADA requires a dedicated database management system as the past data are critical. The accessing of data on any query quickly is achieved by established databases and could have custom design as per the requirement of the master station.
3. *Reporting and Accounting*: Power system hierarchical setup requires a large number of reports and accounts to be prepared for submission to different agencies and also for internal purposes. Hence, this function is very important in power applications. The reporting and accounting software can be predefined but should be customizable to meet future requirements.
4. *HMI Functions*: The software on the console of the operator is of utmost importance, and for the smooth functioning of the control center this software has to be user friendly.

2.8.1.2   Advanced SCADA application functions

This group of application functions is of utmost importance to the power system SCADA because this includes all the basic power system analysis tools required for the proper monitoring and control of the power system. The advanced application functions for generation, transmission, and distribution SCADA are discussed briefly in Chapter 1 and in detail in the following chapters.

2.8.2   Master station hardware components

The main hardware in a master station will be the computer and server systems used for executing the different tasks to be performed by the master station. The computer servers must be selected based on the requirements of the master station.

2.8.3   Server systems in the master station

The SCADA master station consists of a range of server systems, each dedicated to a specific task. The servers are connected through a high-speed dual redundant LAN. The data from one server can be accessed by another in this client-server environment. The dedicated server systems will have special capabilities and features which make them suitable for a particular application. Some of these features are faster CPUs, increased high-performance RAM, redundancy in power supplies, network connections, and high-performance RAMs.

The computer server systems available in a SCADA master station are as follows:

1. SCADA server
2. Application server
3. Information Storage and Retrieval and Historical Information Management (ISR/HIM)
4. Development server
5. Network management server (NMS)
6. Video projection system (mimic board)
7. CFE (communication front end)/FEP (front-end processor): part of I/O system
8. Inter-Control Center Communications Protocol (ICCP)
9. Dispatcher Training Simulater (DTS) server

The main functions of each server are described below.

2.8.3.1   SCADA server

The SCADA server is responsible for all the basic SCADA functions of data collection and display and control command execution from the master station.

2.8.3.2   Application server

The application server hosts the application software modules required for the specific SCADA system. For the generation SCADA system, the application software will include automatic generation control, economic load dispatch, unit commitment, short-term load forecasting, and so on. The transmission SCADA will have the energy management systems (EMSs) package, which includes network configuration/topology processor, state estimation, contingency analysis, three-phase balanced operator power flow, optimal power flow, and so on. Distribution SCADA will have voltage reduction, load management, power factor control, two-way distribution communications, short-term load forecasting, fault identification, fault isolation, service restoration, interface to intelligent electronic devices (IEDs), three-phase unbalanced operator power flow, interface to/integration with automated mapping/facilities management (AM/FM), interface to customer information system (CIS), trouble call/outage management, and so on.

The application software may reside in the SCADA server itself for smaller systems.

2.8.3.3   ISR or HIM server

An information storage and retrieval or a historical information management server supports the reporting accounting activities, and archiving of data for the system. The functions may include real-time data snapshot, historical information recording, retrieval, and report generation.

2.8.3.4   Development server

The development server handles all the initial engineering and commissioning of the project and any changes/developments during the day-to-day running of the system. It is also called the program development system (PDS). This capability is to develop new programs for the system and to implement them. Some of the examples will be software development, display development, and database generation.

2.8.3.5   Network management server

Modern control centers have many digital devices connected via the local area network of the master station. The network management console monitors and manages all the hardware equipment connected to the LAN.

2.8.3.6   Video projection system

A video projection system drives the mimic board displays in a large master station. Master stations are equipped with state-of-the-art LCD display systems that can display the area of control in a varied manner as per the requirement of the operators, and a separate video projection system handles this function.

2.8.3.7   CFE (communication front end) and FEP (front-end processor)

Communication front end (CFE) interfaces the host computer to a network or peripheral devices. CFE is used to offload the host computer of the communication functions such as managing the peripheral devices, transmitting and receiving messages, packet assembly and disassembly, and error detection and error correction. The CFE, often referred to as the FEP (front-end processor) communicates with the peripheral devices using slow serial interfaces, usually through communication networks. FEP communicates with the host computer using a high-speed parallel interface. FEP/CFE is synonymous with the communication controller/ RTU controller discussed earlier.

2.8.3.8   ICCP server

The inter-control center protocol server supports the data transmission between the master and higher hierarchy. This server will also support the inter-site exchange of data with the lower hierarchy. As an example, the ICCP server at a Regional Load Dispatch Center (RLDC) will exchange data between RLDC and National Load Dispatch Center (NLDC) as well as with the State Load Dispatch Center (SLDC) and sub-LDC if required.

2.8.3.9 Dispatcher training simulator (DTS) server

A dispatcher training simulator (DTS) server is generally available at a large master station (at the regional level generally) and is used to train the power system operators who manage the system. The DTS generally provides the power system model, hydro system model, control center model, and instructor functions and is detailed in Section 5.10.

2.8.4 Small, medium, and large master stations

The sub-load dispatch center (LDC) is a small master station (Figure 2.15) with a few RTUs reporting via the CFE. The system will have a console for display and some local control if necessary. The routers for sending the data to the higher hierarchy and peripheral devices will complete the components of the sub-LDC master station.

A medium master station (Figure 2.16) will have many servers, the SCADA/EMS server, development server, ISR server for information storage and retrieval, network management server, inter-control center protocol server, and a large number of operator terminals for monitoring and control. The data come from the field via RTUs/IEDs. Peripheral devices complete the medium master station components.

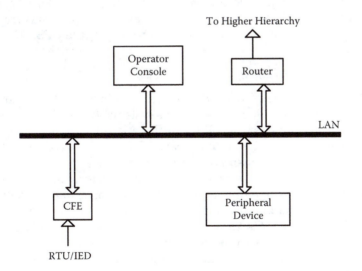

Figure 2.15 Block diagram of a small master station.

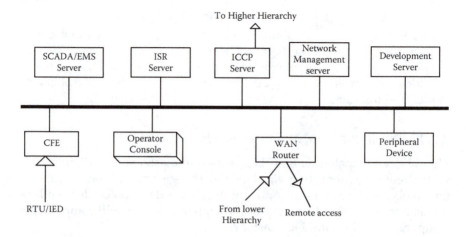

Figure 2.16 A medium master station.

Figure 2.17 shows a large master station with a dual redundant LAN, with all the components of the medium master station, a few additional servers, and additional layers of security. The large master station will generally have a complete redundant station at a remote location, so that any natural calamity affecting one station will not cause a problem with the functioning of other stations and the system can be monitored and controlled effectively. Uninterrupted power supplies (UPS) monitoring systems are also of importance, as all the processors and servers in the master station are fed from the substation battery by the UPS.

2.8.5   *Global positioning systems (GPS)*

All SCADA components from RTU to IEDs to master stations are time synchronized by the global positioning systems' (GPS) clocks. The GPS has 24 to 32 geostationary satellites in the mid orbit of the earth, which continuously transmit signals to the earth. These signals contain information about the time of transmission and the satellite location at the time of transmission. Signals are received by the GPS receivers at different locations on the earth. The receivers compute the distance to the satellite and by using some equations, the location of the receiver and the time can be accurately monitored. The receiver should accommodate four or more visible satellite data signals to correctly identify the position. The GPS space and control segments are developed and controlled by the US military and are used by millions of devices across the world.

In power system SCADA, GPS plays a vital role, as time synchronization is the key for all measurements, especially IED measurements, including phasor measurement units.

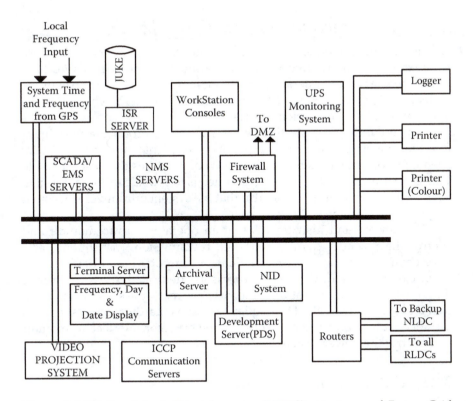

Figure 2.17 National load dispatch center (NLDC) structure of Power Grid Corporation of India Limited (PGCIL) (large master station). (Courtesy of Power System Operations Corporation, India.)

2.8.6 Master station performance

The scenarios are typically defined for two activity levels. They are the normal state and a peak activity state. Each activity state is defined by percent of each type of point changing behavior scans, applications that are running and frequency of their execution, different displays called up at certain frequencies, and so on. The scenarios are conducted for specified times, usually between 10 and 30 min. At the end of each test, performance parameters such as CPU utilization and LAN utilization for each CPU and LAN are tabulated and compared to the required levels. The results of these tests determine whether the system is acceptable. If the system is not acceptable, the supplier is presented with quantitative results so that the problem can be reviewed and corrected. The system is retested, when necessary, to demonstrate that the corrective action was successful.

2.9   Human-machine interface (HMI) [24,25]

Human-machine interface (HMI) or user interface (UI) refers to the space where the interaction between the humans and the system (machine) happens. The goal of this interaction is effective operation and control of the system being monitored, and feedback from the system, which aids the operator in making operational decisions.

Generally, the goal of HMI engineering is to produce a user interface that makes it easy or self-exploratory, efficient, enjoyable, and user friendly to operate a system in the way that produces the desired result. This generally means that the operator needs to provide minimal input to achieve the desired output, and also that the system minimizes undesired outputs to the human.

The devices and instruments that an operator uses to monitor and control the power system (generation, transmission, distribution, etc.) have changed drastically from manual to computer-based devices. The latest expensive hardware, fast processors, dedicated software, mimic diagrams, and communication protocols have made the system very compact and human friendly. The display functions (hardware and software) installed in the control center serve as the operator interface to monitor and control the power system.

2.9.1   HMI components

In a SCADA system, the HMI components include operator console, operator dialogue, mimic diagram and peripheral devices.

2.9.1.1   Operator console

The console where the operator monitors and controls the system is of utmost importance and includes the visual display units, alphanumeric keyboard, cursor, communication facilities, and so on. The visual display unit includes UI devices like the multiple color monitors (CRT, LCD, LED devices minimum 21″ size), with glare-reduction features (antiglare screen coatings) and should provide a display of multiple viewports (windows) on each monitor. The cursor control could be mouse, trackball, or the latest touch-screen facility. A keyboard and cursor pointing device are shared among all monitors at each console and the cursor moves across all screens without switching by user. Generally for power system SCADA each operator has three to four monitors for proper planning and multiple views. The displays will have full graphics capability with zoom and decluttering facility. Audible alarms are also a prominent feature of the operator console where the operator is informed of the severity of an event in the system. The design of the operator console infrastructure including the table and chair for the operator is important and should follow ergonomics principles to make the operator comfortable during the duty period.

2.9.1.2   Operator dialogue

Operator dialogue is how the operator communicates with the computer system. The operator dialogue and commands should be simple and easy to remember. Function keys of the keyboard can be programmed to incorporate major actions so that the operator can give the commands easily rather than typing long messages and dialogues.

2.9.1.3   Mimic diagram

The mimic diagram is an essential part of any control center or large master station where the operator and the personnel in charge get an overall view of the system under control. This includes LCD/LED large-screen display with full SCADA operability with multiple screens possible. Some control centers have mosaic map board with dynamic or static tile map board and dynamic map board lamps updated by SCADA. Magnetic map boards with static magnetic "tiles" are also in use. The trend is to use multiple video projection "cubes" as the dynamic map boards. The advantage is that when the HMI is updated with system changes, the map board is also updated, since the HMI drives the map board directly.

2.9.1.4   Peripheral devices

A dot-matrix printer is used to print alarms and events. It uses fanfold computer paper. A color printer is used for capturing screen shots. A black and white laser printer is used to print reports.

2.9.2   HMI software functionalities

- *Access Control Mechanisms*: The operator console should have levels of security the ability to protect unauthorized access to the system. Specific user identification (IDs) and password will be used by authorized system operators to gain access to the operator consoles.
- *Visualization and Control*: The visual presentation of power system information and to control the same is an effective way of presentation and increases the efficiency of the operator. The video display unit at the control center displays all the information about the power system interconnection and the parameters of interest, viz. voltage, current, frequency, power flows through the tie lines and the connected areas, which the operator monitors and uses to analyze the events and to take corrective and control action, if necessary.
- *Standard System Displays*: Diagnostic, site and industry-specific displays, graphical displays with drill-down facility, and a display hierarchy are important. Also pan and zoom facility, decluttering,

and layering for better clarity of problems and possible solutions are musts for modern display systems. Easy navigation through the display is possible with the cursor and control device.

- *Historical Trending, Trending Display, Real-Time Trending*: The operator estimates the data from the past history and the information about the process variables, which are helpful in predicting the future state of the system and its health. The control center provides the history of the information received with trending to the operator. Fast-acting computers with large memory are able to store the power system variables in the database and provide the information to the operator and higher hierarchy in minimum time with greater accuracy. Real-time trending is also very useful in presenting the correct picture of the system variables to the operator.

- *Logs and Reports, Calculated Values, Spreadsheet Report Generation, Reports as Data Exchange Mechanism*: In power system SCADA applications, report generation is a major task as many types of reports have to be generated for presentation to various system hierarchies and to different departments of the utility. With the increased information available with modern SCADA systems, data warehousing is a must and will be dealt with separately in this book in Section 4.10.

- *Alarm Processing*: Once the data are communicated to the control center, they are first processed before presentation to the operator. The processed data are then compared with the predefined values and in case of any deviation from the nominal value, an alarm is generated, the operator acknowledges that and necessary corrective action is taken. Thus, important functions of the control center allow it to generate, annunciate, and manipulate the process and system alarms. The generation and display of the alarm and the limits are important functions of the control center, and this information needs to be communicated to the interconnected system in case of emergency, limit violations, and malfunctions. Sometimes, the operator is confused by the series of alarms triggered by a single event, as many quantities change and unnecessary alarms of all kinds are directed to the operator. Hence, alarm filtering is important and is explained in the following section.

2.9.3   Situational awareness [9]

Situation awareness in general is knowing what is going on around you, so that you can decide what to do. The operator in a control room should have a perception of the environment around him or her in time and/or space and comprehend the meaning and should be able to make decisions on future actions, depending on the situation. Situational awareness is used in aviation, air traffic control, ship navigation, military operations, and emergency services like firefighting and in power system control rooms now. Situational awareness is important when the information flow is fast and the error in judgment will lead to major consequences, for example in power systems, a blackout if appropriate action is not taken on time.

Many researchers have defined situational awareness, the most popular one being "the perception of elements in the environment within a volume of time and space, the comprehension of their meaning, and the projection of their status in the near future" [10]. The model developed includes three levels: perception level, where the person has to perceive the status, attributes, and dynamics of the variables in the environment; comprehension level, where the data from level one have to be synthesized using interpretation, pattern recognition, and evaluation skills; and the projection level, where the person can extrapolate the information from the lower levels and arrive at an action plan.

Operators make errors when they are not completely aware of the ground situation, so the aim is to equip the control room with visualization aids for level one, have enough data synthesizing and display systems for level two, and finally enable the operator to take a decision and implement it.

SCADA master station software visualization has undergone a major transformation with new devices and tools available for increasing the perception and comprehension levels of the operators, to equip them to make better decisions at the projection level (discussed in detail in Section 5.12).

2.9.4   Intelligent alarm filtering: Need and technique [7]

Alarm-processing technology ensures that dispatchers receive only those alerts relating to events that must be addressed immediately, while the details of less critical secondary warnings are sent to databases and possibly printed for later review. With only the most important distribution system alarms presented in a prioritized fashion, dispatchers can assess problems more easily and make better decisions to prevent a bad situation from getting worse.

The reason that alarm-processing technology has been implemented early enough in distribution SCADA (SCADA/DA/DMS) and not in transmission SCADA (SCADA/EMS), where it is being implemented now, is a combination of application necessity and customer demand. And the fact that companies providing distribution SCADA products are typically different from those offering SCADA/EMS has not helped the situation. Fortunately, these two types of SCADA systems operate similarly, which means distribution alarm suppression technology can readily be implemented in SCADA/EMS.

On the distribution side, SCADA alarms are typically triggered by faults and the events surrounding them, which occur continuously during routine operations. When a breaker on a substation feeder trips due to a transient fault, for instance, up to seven alarms may be triggered: one for the breaker trip and three each when voltages and currents on all three phases hit zero. The dispatcher needs only the breaker trip alarm and does not need any alarm information if the breaker is automatically reclosed after a transient fault since the situation resolves itself.

With audible and visual alarms inundating the control room throughout the day, dispatchers asked distribution SCADA vendors to suppress secondary alarms while allowing primary alarms, though primary alarms require operator action while secondary alarms need no operator action. In response to this demand, the vendors developed filtering techniques, some of which can be configured during SCADA implementation or activated during a storm.

Such demand never occurred on the generation and the transmission sides since SCADA/EMS alarms are triggered less frequently and only during actual outage events. Because these alarms have not posed the same daily nuisances, utilities simply never pressured vendors to implement alarm filters in SCADA/EMS until now.

2.9.5 Alarm suppression techniques

The 2003 blackout has compelled the power industry to revisit the alarm issue. With prompting from utility customers, SCADA/EMS vendors are incorporating filtering techniques into the control center software. In general, there are four proven alarm-processing methods currently used in distribution SCADA systems and vendors can choose them for implementation in their future products.

2.9.5.1 Area of responsibility (AOR) alarm filtering

Inherent in the SCADA architecture is the ability to partition the system by function or geography. This allows a utility to separate the monitoring and operation of various SCADA displays, alarms, and control points and assign responsibility for them to different control rooms, dispatchers, or even other utilities. Distribution SCADA systems can usually be partitioned into 64 functions or geographic areas.

SCADA systems are designed this way due to the broad variety of their application. A water, gas, and electric utility, for instance, may want to invest in only one SCADA system but establish separate control rooms for each of its three services. Creating AORs for gas, water, and electric services accomplishes this. A more common example is a generation-transmission co-op that turns distribution network monitoring functions over to its electric member cooperatives from 9 a.m. to 5 p.m. every day and resumes those functions overnight and on weekends.

AOR partitioning also gives utilities tremendous flexibility in routing alarms. For example, all operational alarms can be sent to the control room, while equipment monitoring alarms go to maintenance. Or one dispatcher may receive alarms pertaining to one geographic region while another dispatcher gets the alarms for another geographic region. The variations of alarm partitioning are almost endless, but the bottom line is that this enables the utility to filter the alarms so that only the most important reach the people who can handle them.

In SCADA/EMS systems already installed in the past few years, partitioning may already be built into the system, although the number of partitions may not be as numerous as those in distribution SCADA. Regardless, power system alarms can be divided and filtered in the same way as those on the other side and with the result being less distraction for the dispatcher.

2.9.5.2   Alarm point priority filtering

During the configuration of a distribution SCADA database, each monitoring and control point in the network is assigned an alarm priority level by the utility. These points usually rank in importance from one to eight, with eight being the most critical. On the distribution side, for example, breakers on critical feeders could be assigned high numbers.

At console in the control room, dispatchers can select which alarms they want coming through to their display windows based on priority level. In daily operations, the dispatcher may want to see alarms from all priority levels on the screen, but when a storm starts moving into the territory, for example, the dispatcher can dynamically change the preference to show only priority alarms six and higher. This gives dispatchers control over the filtering and suppression of warning based on the gravity of the situation at hand.

2.9.5.3   Timed alarm suppression

When a SCADA system is configured during installation, the utility can determine the length of time that an out-of-threshold situation must last before it actually triggers an alarm. If the situation is transient or resolved before this time period elapses, the trigger never occurs, and the dispatcher is not bothered with a noncritical event, although details are still written to the alarm and event disk file and possibly printed.

An example is a distribution feeder with an automatic recloser. When a tree branch blows against the feeder in the midst of a wind storm, the recloser opens and then recloses as programmed. If the branch is no longer striking the line, the recloser remains closed. But the dispatcher would needlessly receive two alarms, one for the opening and one for the reclosing, despite the fact that normal operations had been restored.

Under the timed-alarm suppression process, a timer begins when the recloser first opens, and no alarm is activated. Once the predetermined time period ends, perhaps 2 to 4 s, the SCADA system again looks at that point to see if the recloser is still open. If it is, the alarm is triggered and the dispatcher knows a situation more serious than a transient condition has occurred. Otherwise, if the SCADA system finds the feeder operation has returned to normal, there is no alarm.

For use in generation and transmission SCADA operations, the timed suppression technique would be applied to the status of transmission lines and power generation units. Since SCADA/EMS typically monitors whether these components are inside or outside certain limits, acceptable durations of threshold exceptions can easily be assigned to each control point for alarm suppression.

2.9.5.4 Knowledge-based alarm suppression

Within the SCADA database, direct linkages can be created between network elements that trigger primary and secondary alarms. By linking them, the secondary alarms can be eliminated if the primary one has already been activated. This can be illustrated using the above example of the feeder opening that causes voltage and currents to drop, sending six needless alarms to the dispatcher.

Database records are created for each voltage and current point on the feeder and linked to the status of the feeder breaker in the database. If the value drops to zero at any of those points, triggering a low-burst alarm, an address pointer in the SCADA database will automatically check the measurement point of the feeder breaker status before activating the low voltage or current alarm. If the breaker status is open, the SCADA system knows a primary alarm has already been triggered, and it suppresses the redundant low voltage alarm. Occurring in a split second, this process then records the secondary alarm in the alarm and event disk file and sends the alarms to the alarm and event printer.

Feeder and breakers are elements of the electric distribution system, but knowledge-based alarm suppression can be applied just as easily on the generation and transmission side. The key in implementing this technique in SCADA/EMS is identifying and linking critical system functions that secondarily impact other operations that can trigger alarms. For example, when a generator breaker trips, the terminal voltage will go to zero. The breaker trip would be the primary alarm, and the terminal voltage would be a secondary alarm.

SCADA vendors are implementing one or more of these filtering and suppression techniques. Since the basic technology that makes AOR and alarm point filtering possible already exists in some SCADA/EMS, it is likely these two will emerge as the dominant alarm-processing methods.

2.9.6 Operator needs and requirements [25]

The master station is the location from where the operator/system engineer monitors the health of the power system and issues necessary control instructions. This is where the operator typically spends 6 to 8 working hours monitoring and controlling the system. Proper design of the control center should ensure ease of functioning for the operator, without exerting stress in vital body parts. Suitable atmosphere and housing have to be created and the operator console design has to adhere to standards so that the operator is comfortable and can concentrate on the job.

Some of the needs and requirements of the operator in the control room, to meet the functional objectives, are listed as follows:

1. *Flexibility of Workspace*: The operator working in the control center for long durations to handle functional requirements must be able to monitor and control the operations of the power system equipment via the various input and output devices connected at the console. The operator should move from one device to another with ease.
2. *Ease of Operation of Control Devices*: The operator uses the user-friendly operation and control system.
3. *Ease with Which the Operator Reaches the Control Point*: The operating and control system, which the operator is using, ensures that the alarms and indicating displays are easy to locate at the time of emergency and necessary action is taken when required.
4. *Good Resolution with Least Error*: The display units display the mimic and the other related information that is easily readable and also provides accurate and proper information to the operator.
5. *Training Requirement*: The proper training of the operator in operation and control is required from time to time.
6. *Eye-hand Coordination*: The operator console, the input devices, the display units, and so forth, must provide a comfortable approach for handling the equipment.
7. *Pleasant Atmosphere*: The operator working for long durations requires other facilities, viz., entertainment, exercise, yoga, tea or coffee, time for rest, and so forth in an operator-friendly control center.
8. *Less Fatigue*: Comfort in the use of devices, stability and reliability of the equipment, proper training, and so on, ensures reduction of stresses in different parts of the body and improves working efficiency of the operator.

2.10 Building the SCADA systems, legacy, hybrid, and new systems

The above sections elaborated the building blocks of SCADA systems, starting from the RTU, IEDs, communication systems, master stations and the HMI. Utilities have a variety of options available to mix and match the elements to build a cost-effective, efficient, and operator-friendly SCADA system.

However, automation of the power systems started as early as the beginning of the twentieth century, and substations and control centers operate at various stages of automation all over the world. There are legacy systems with RTUs, hardwired communication from the field to the RTU, and traditional software functionalities in the control room, and it is not often financially viable to dismantle everything and purchase a completely new automation system.

Hybrid systems are a viable option, where any automation expansion project can be implemented with new devices, like IEDs, data concentrators, and merging units, as shown in Figure 2.14. The new system will coexist with the legacy RTU-based systems, and the data integration and if necessary protocol conversion issues will have to be handled while commissioning the project.

If a utility decides to purchase a completely modern system, the latest building blocks of the SCADA system, viz., IEDs, merging units, and fiber optic communication facility with brand new HMI with situational awareness and analysis tools, can be implemented.

The legacy, hybrid, and new systems for a typical substation automation implementation are discussed in detail in Section 4.7.

2.11 Classification of SCADA systems

SCADA systems can be classified into four categories depending on the complexity and the number of RTUs and master stations present in the system. The classification will also depend on the number of points at each RTU and the required update rates, location of the RTUs, communication facilities, and equipment available.

2.11.1 Single master–single remote

The simplest configuration, the single master–single remote supervisory (Figure 2.18), is utilized for simple systems where small numbers of points are involved, since it requires one master station and one communication channel per RTU. This one-on-one configuration generally has one indicator or display at the master station for each remote data point. An

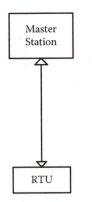

Figure 2.18 Single master–single remote configuration.

example could be the control center of a generating station with one RTU to collect data.

2.11.2 Single master–multiple RTU

In the one master–multiple RTU configuration, one master station is shared by several RTUs. Generally, the master station communicates in turn to each RTU using serial digital data messages. This configuration has the advantage over the one-on-one of sharing the master station communications logic among a number of RTUs. An example could be a power distribution system with one master station controlling a number of substations with RTUs. These are generally "off-the-shelf" systems that can be procured easily, and the number of RTUs is generally restricted to 25. The communication configuration could be radial or shared line (party line) as shown in Figures 2.19a and 2.19b.

2.11.3 Multiple master–multiple RTUs

In multiple master–multiple RTU configuration, there will be submasters available with multiple RTUs reporting to each master. These systems will have a large number of RTUs connected to it, and extensive engineering and customization are required for commissioning of the system. The multiple master–multiple remote is also characterized by a sizable number of application programs. These systems will take a longer time to execute and implement. Figure 2.20 gives the multiple master–multiple RTU concepts. An example is a generation and transmission (G&T) utility with multiple distribution members, where each member has its own SCADA system. The member's SCADA masters report upstream to the G&T master.

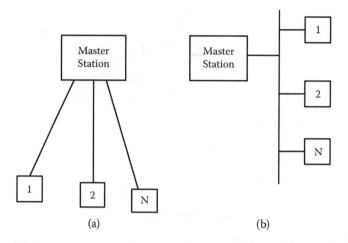

Figure 2.19 (a) Single master–multiple RTU (radial). (b) Single master–multiple RTU (shared line).

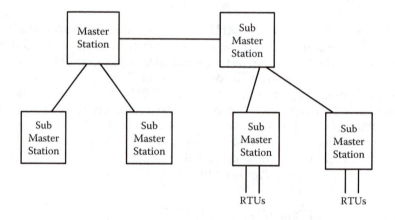

Figure 2.20 Multiple master–multiple remote configuration.

2.11.4   *Single master, multiple submaster, multiple remote*

In this system there is a single master, with additional submasters, with each submaster reporting to the master station. The remote RTUs/IEDs will typically be connected to the submasters. A typical system would be the hierarchical transmission SCADA used. Figure 2.21 gives the practical system, which is the National Control Center for the transmission SCADA in India. The multiple submasters represent the five regional control centers and also 29 state load dispatch centers. The RTUs are located in the substations around the country.

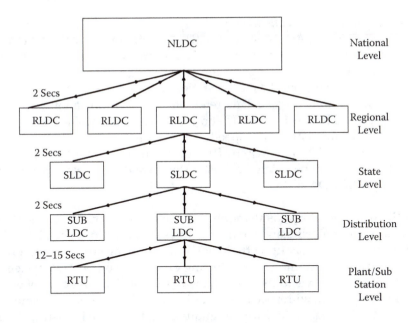

Figure 2.21 Hierarchical setup of the Indian transmission system. (Courtesy of POSOCO.)

2.12   SCADA implementation: A laboratory model [2,5,16]

The SCADA components are clearly depicted in a laboratory implementation where undergraduate and graduate students undergo a complete commissioning process and gain hands-on training on a real SCADA system, as explained in the following section.

2.12.1   The SCADA laboratory

SCADA systems are used worldwide in a variety of automation applications in the gas and petroleum industry, power automation, building automation, and small manufacturing unit automation. SCADA systems, though used extensively by the industries, are proprietary to each company, and hence, very few technical details are available to students and researchers. The setting up of the SCADA laboratory in India was of great relevance as this provides research facilities in the form of hardware and software for adaptive and intelligent control of integrated power systems.

SCADA systems have the following four components:

1. *Master Station*: This is a collection of computers, peripherals, and appropriate I/O systems that enable the operators to monitor the state of the power system (or a process) and control it.

2. *Remote Terminal Unit (RTU)*: The RTU acts as the "eyes, ears, and hands" of a SCADA system. The RTU acquires all the field data from different field devices, processes it, and transmits the relevant data to the master station. At the same time, it distributes the control signals received from the master station to the field devices.
3. *Communication System*: This refers to the communication channels employed between the RTU and the master station. The bandwidth of the channel limits the speed of communication.
4. *Human-Machine Interface (HMI)*: HMI refers to the interface required for exchanges between the master station and the operators or users of the SCADA system.

The SCADA laboratory has all the above components of the SCADA system with online monitoring and control facilities as shown in Figure 2.22. The laboratory has been set up with the view of providing students and practicing engineers with hands-on learning experience on SCADA systems, and their applications to the management, supervision, and control of an automated system, with special emphasis on an electric power system.

The SCADA laboratory uses a unique combination of industrial automation hardware and software, integrated with field equipment related to power systems. The lab has been in use for the past 10 years for research and training in industrial automation, as well as power automation. One of the unique features of the SCADA laboratory that makes it the only one of its kind is the use of a distributed processing system which supports a global database.

2.12.2   System hardware

The master station has 12 engineering consoles for project implementation and an equal number of operator consoles for system monitoring. The SCADA hardware includes a distributed processing unit (DPU), a RTU, and a number of analog, digital, and pulse I/O units and field equipment.

The DPU is configured around a 32-bit restricted instruction set computer (RISC) processor AC800F, as shown in Figure 2.23. It can support up to 100 master-less RTUs. At present there is only a single RTU communicating to the DPU. The DPU has the capability of handling more than 1000 inputs and outputs, and is presently configured for 216 inputs and outputs (digital, analog, and pulse). The RTU, DPU, and I/O units are interconnected through the Profibus module. The DPU has the Modbus module for dedicated communication with IEDs, an energy analyzer in this case. The laboratory incorporates industry standard networking. It has an Ethernet data highway operating at 10 Mbps and is currently

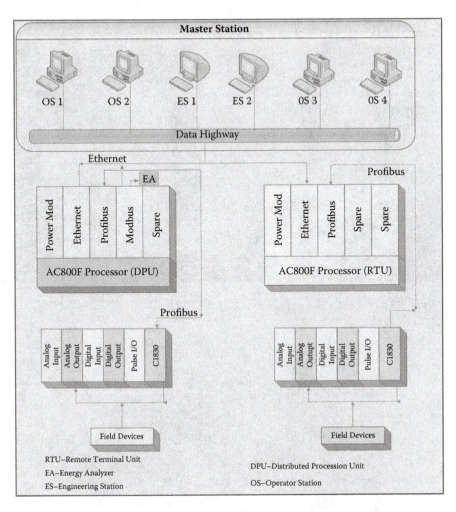

Figure 2.22 Overview of the SCADA laboratory.

supporting a network of 10 operator stations and two engineering stations along with the DPU and the RTU, all connected in bus topology. Figure 2.24 shows the DPU and RTU as set up in the lab.

2.12.3 System software

The laboratory utilizes two system software programs for better understanding and proper utilization of the product available in the market. The first one is hardware-specific with dedicated software, Freelance 2000, whereas the other is an open-ended system software, SCADA portal,

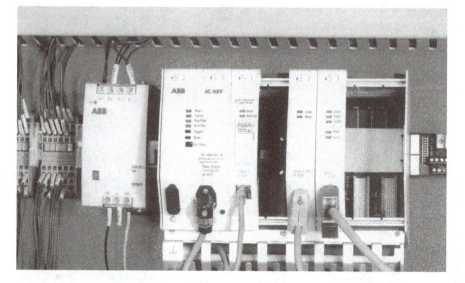

Figure 2.23 (See color insert.) Processor of the DPU.

Figure 2.24 The DPU, RTU with I/O units.

that can communicate with any hardware device. This software is capable of supporting standard power system software programs. The system software has the facility for easy online configuration for mimics, trends, reports, and so on, and for Web navigation. The software relational diagram is given in Figure 2.25.

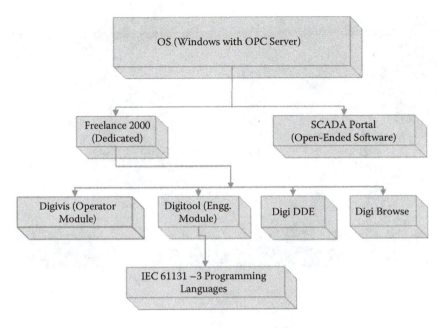

Figure 2.25 SCADA laboratory software relational diagram.

2.12.4   SCADA lab field design

The foremost task in the designing of the laboratory was defining the power system to be monitored and controlled. This was done taking into account adequate scope for expansion of the system in future, and the latest facilities available in instrumentation and monitoring areas. A number of big industrial houses involved in power system SCADA were contacted, and detailed discussions were carried out with the experts in the field. Also a study of the available industrial SCADA systems was done. Finally, the power system to be monitored, the configuration of the laboratory, and the specifications for the field device were finalized. The laboratory field presently includes the following:

- An 11 kV substation feeds the Faculty of Engineering building, Jamia Millia Islamia. The voltages and currents as analog signals and the capacitor bank positions as digital signals are monitored from the 11 kV substation.
- A three-phase transmission line model is used, complete with reactive and capacitive compensation. The transmission line model, as shown in Figure 2.26, was built to simulate the real-time transmission condition in the laboratory so that students can get a hands-on

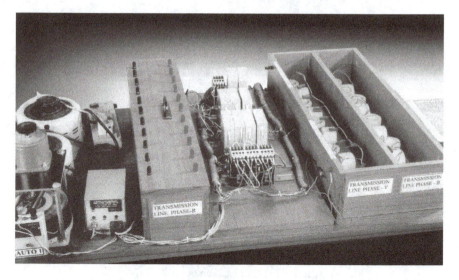

Figure 2.26 (See color insert.) Three-phase transmission line model with OLTC, isolators, and transducers.

experience of the monitoring of the voltage and current (through transducers), circuit breaker closing and opening (through the isolator), and phenomena such as Ferranti effect, series, and shunt compensation, by performing experiments on the system.

- A prototype model of an on-load trap changer (OLTC) using stepper motor and autotransformer is presented. The transmission line is fed from autotransformers with the input voltage varied using a stepper motor, depicting an OLTC transformer.
- An energy analyzer IED gives a variety of digital output signals via Modbus to the DPU.
- RTD, level sensors, transducers, contactors, and so forth, are used to depict the industrial automation scenario.

The undergraduate and graduate students undergo the complete commissioning process including physical wiring, tag allocation, I/O mapping, software customization, and graphic design for a specific process. The SCADA laboratory thus equips the students to take up the challenges in the automation job market at once, and helps the students find suitable placements in global automation companies.

2.13   Case studies in SCADA

The following case studies establish the facts discussed earlier in the chapter by discussing some practical implementations of SCADA projects across the world, where legacy systems and hybrid systems have been established.

2.13.1 "Kentucky utility fires up its first SCADA system" [20]

The Frankfort Electric and Water Plant Board (FEWPB) in Kentucky lost one substation to fire due to an internal fault in a single-phase regulator in the late 1990s, and the utility realized the value of a SCADA system for early warning and timely action. In early 2000, the utility started the implementation of a SCADA and substation automation project for the entire system and successfully implemented it, integrating the old systems with the new devices and a single-user interface.

2.13.2 "Ketchikan Public Utilities finds solutions to outdated, proprietary RTUs" [21]

Ketchikan Public Utilities in Alaska faced serious problems with proprietary RTUs and communication protocols which left it in the dark after the vendor who supplied the master station and RTUs went out of business. A joint effort by the team of engineers from the company and vendors saved the system with a new master station in the control room and protocol converters in the substations.

2.13.3 "Overwhelmed by alarms: The blackout puts filtering and suppression technologies in the spotlight" [22]

The blackout in 2003 in the United States emphasized the importance of intelligent alarm processing as the operators could not comprehend the stream of alarms that were generated during cascading events during an emergency. Alarm-suppression techniques are discussed in detail to inform the operator of significant events only, the primary alarms, so appropriate action can be taken and secondary alarms will not be sent to the operator's alarm summary. (More details are presented in Section 2.9.4.)

2.13.4 "North Carolina Municipal Power Agency boosts revenue by replacing SCADA" [23]

The North Carolina Municipal Power Agency had a SCADA system, with legacy RTUs and frame relay communication, which also handled the meter data from 47 substations for revenue computations. However, the utility could not take advantage of the power marketing by block scheduling, due to the forecasting error, as the RTUs were polled every 5 min, causing inaccuracy. With a new SCADA system and proper analysis of the communication system, the scan time was brought down to 4 s, which was enough for the utility purposes.

2.14   Summary

This chapter gives a comprehensive description of the components of SCADA systems which include RTUs, IEDs, data concentrators/merging units, master stations, and the HMI. Communication systems and protocols play a major role in power system SCADA as the field is spread over a large area and fast communication is required due to the dynamic nature of the power system. The next chapter explains the SCADA communication and the protocols used in power system SCADA.

References

1. J. D. McDonald, Substation automation, IED integration and availability of information, *IEEE Power & Energy Magazine*, vol. 1, no. 2, pp. 22–31, March/April 2003.
2. Mini S. Thomas, Remote control, *IEEE Power & Energy Magazine*, vol. 8, no. 4, pp. 53–60, July/August 2010.
3. Stuart A Boyer, *SCADA Supervisory Control and Data Acquisition*, IVth ed., International Society of Automation (ISA), Research Triangle Park, NC, 2010.
4. William T. Shaw, *Cyber Security of SCADA Systems*, Pennwell, Tulsa, OK, 2006.
5. Mini S. Thomas, Pramod Kumar, and V. K. Chandna, Design, development and commissioning of a supervisory control and data acquisition (SCADA) laboratory for research and training, *IEEE Transactions on Power Systems*, vol. 20, pp. 1582–1588, August 2004.
6. John D. McDonald, *Electric Power Substation Engineering*, 3rd ed., CRC Press, Boca Raton, FL, 2012.
7. John D. McDonald, Overwhelmed by alarms, blackout puts filtering and suppression technologies in the spotlight, *Electricity Today*, no. 8, 2003.
8. Mini S. Thomas, D. P. Kothari, and Anupama Prakash, Design development and commissioning of a substation automation laboratory to enhance learning, *IEEE Transactions on Education*, vol. 54, no. 2, pp. 286–293, May 2011.
9. J. Giri, M. Parashar, J. Trehern, and V. Madani, The situation room: Control center analytics for enhanced situational awareness, *Power and Energy Magazine, IEEE*, vol. 10, no. 5, pp. 24–39, 2012.
10. M. R. Endsley, Toward a theory of situation awareness in dynamic systems, *Human Factors*, vol. 37, no. 1, pp. 32–64, 1995.
11. Mini S. Thomas, D. P. Kothari, and Anupama Prakash, IED models for data generation in a transmission substation, *Proceedings of the IEEE Conference: PEDES-2010*, New Delhi, India. DOI: 10.1109/PEDES.2010.5712415.
12. T. Sezi and B. K. Duncan, New intelligent electronic devices change the structure of power distribution systems, *Industry Application Conference, 1999, 34th IAS Annual Meeting*, vol. 2, pp. 944–952, October 3–7, 1999.

13. P. G. McLaren, G. W. Swift, A. Neufeld, Z. Zhang, E. Dirks, and M. Haywood, Open system relaying, *IEEE Transactions on Power Delivery*, vol. 9, no. 3, July 1994.
14. Ching-Lai Hor and Petter A. Crossley. Extracting knowledge from substations for decision support, *IEEE Transactions on Power Delivery*, vol. 2, no. 2, part I, 2005, pp. 595–600.
15. Cobus Strauss, *Practical Electric Network Automation and Communication Systems*, Newnes, Elsevier, Amsterdam, The Netherlands, 2003.
16. Mini S. Thomas and A. K. Srivastava, On-line monitoring of SCADA systems: A practical implementation, *IEEE International Conference: PEDES*, New Delhi, India, December 2006.
17. Anupama Prakash, Mini S. Thomas, and Ashutosh Gautam, Integration of IEDs using Legacy and IEC 61850 protocol, *IEEE International Conference: PEDES*, New Delhi, India, December 2006.
18. Parmod Kumar, Vinay Chandna, and Mini S. Thomas, Fuzzy genetic algorithm for pre-processing data at RTU, *IEEE Transactions on Power Systems*, vol. 19, no. 2, pp. 718–723, May 2004.
19. Parmod Kumar, Vinay Chandna, and Mini S. Thomas, Intelligent algorithm for pre-processing multiple data at RTU, *IEEE Transactions on Power Systems*, vol. 18, no. 4, pp. 1566–1572, November 2003.
20. Dave Carpenter, Vent Foster, and John D. McDonald, Kentucky utility fires up its first SCADA system, *T&D World*, February 2005.
21. Harvey Hansen and John D. McDonald, Ketchikan Public Utilities finds solutions to outdated, proprietary RTUs, *Electricity Today*, no. 2, 2004.
22. John D. McDonald, Overwhelmed by alarms: The blackout puts filtering and suppression technologies in the spotlight, *Electricity Today*, no. 8, 2003.
23. John D. McDonald, North Carolina municipal power agency boosts revenue by replacing SCADA, *Electricity Today*, no. 7, 2003.
24. IEEE Tutorial course on Fundamentals of Supervisory Systems, Course 94, EH0392-1PWR.
25. Pavmod Kumar, V. K. Chandna, Mini S. Thomas, Ergonomics in content centre design for power systems, IEEE Power India Conference 2006, New Delhi, India. DOI: 10.1109/POWERI.2006.1632584.

chapter three

SCADA communication

3.1   Introduction

Supervisory control and data acquisition (SCADA) communication refers to the communication channels employed between the field equipment and the master station. The channel makes it possible for a remote control center to access the field data in real time for assessing the state of the system, whether the generation by each unit, the voltage and current vectors from buses, the loading on the system, or circuit breaker and isolator positions. The communication channel also transports the control commands from the control center to the appropriate equipment in the field for implementation, to keep the power system stable and secure. SCADA communication is analogous to the nervous system of the human body which runs from the brain to every part of the body transporting data and signals back and forth continuously.

SCADA communication was previously restricted to major equipment and buses; however, with the deployment of smart grid, two-way communication runs to the end customer as automation extends from generation to transmission, to distribution, and finally to customer, with home automation now implemented in a big way. Communication media, protocols, and deployment are significant.

As compared to industrial automation systems, communication systems are predominant in power automation systems due to two major reasons:

1. The extent of the power system, with control areas spread over large geographical regions, extending to thousands of kilometers, implies that the communications system has to be robust, reliable, and physically feasible.
2. The speed of data transfer required in a power system for critical data is in milliseconds, which makes it imperative to use technologies and protocols that aid in this aspect. A fast communication channel is essential for functions like phasor measurement unit (PMU) applications, automatic generation control, transient stability, and oscillation data, and for demand response mechanisms to work. The opening and closing data for circuit breakers, isolators, and switches should

reach the control room in 1 to 2 s, whereas the analog measurement values must reach control within 15 to 60 s. Other data such as metering, waveform data, and so on, can be acquired at longer intervals.

3.2   SCADA communication requirements

Communication requirements can be defined as those elements that should be considered for proper functioning of the communication system. Some important requirements for SCADA communication are as follows:

- Communication traffic flows must be identified, which include the quantity of data to be transferred, the source of data, and the destination where the data are to be transferred. The identification of end system locations is also important.
- System topology—ring, star, mesh, or hybrid–is important.
- The capabilities of the devices used for communication at both ends and the processor capabilities are noted. The device addressing schemes are also important in design of a communication system.
- The communication session and dialogue characteristics need to be explored during the design phase.
- Communication traffic characteristics are very important due to the time-critical data transfer requirements of the power systems.
- Performance requirements of the communication system must be known.
- Reliability of the communication system, backup system, and failover is critical.
- Timing of the communications is significant.
- Application data format and the application service requirements are important.
- Electromagnetic interference must withstand capability qualifications.
- Operational requirements such as directory, security, and management of the network are important.

The communication requirements of the SCADA systems are stringent, as discussed; however, with smart grid implementation gaining momentum, the related communication requirements also need to be addressed.

3.3   Smart grid communication infrastructure

Distribution systems were minimally automated, and at the most, monitoring was performed at substations for sending the data to higher hierarchies. So when a simple distribution system is strengthened with communication and control capabilities, in addition to monitoring, it migrates to a smart grid. The same communication infrastructure is used to get the consumer inputs and to send information from the smart grid control center to the consumer. The motivation behind developing an effective communication infrastructure in a smart grid is related to improving the system and its operation to benefit the customer and also to protect the environment. [1]

Figure 3.1 shows the communication infrastructure extension to customers, making it possible to reap many benefits. The first benefit of an improved communication infrastructure of a smart grid could be better customer experience. This can be enhanced by fast notification if there is any power supply interruption or fault. The outage time will be reduced, and this makes the service more reliable. The communication infrastructure helps in knowing the customer consumption of electricity in response to supply condition and provides users with some tools to reduce their consumption during peak hours so as to make the power system more reliable. By ensuring customers' participation, productivity increases and maintenance and operation costs will be reduced. System operator performance will also improve as the operator will receive all the crucial real-time information, and thus the decision-making capability for fault isolation or faulty component replacement will be improved.

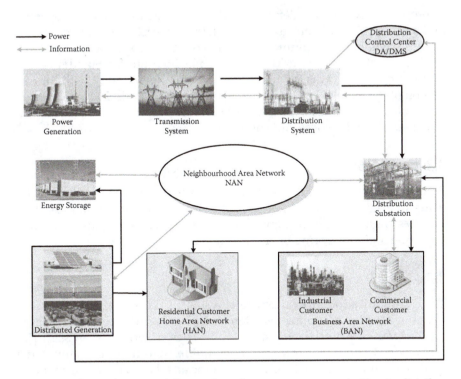

Figure 3.1 (See color insert.) Expansion of two-way communication to distribution system including customers.

Renewable energy generation and integration are necessary to reduce the carbon footprint and greenhouse gas emissions. A smart grid enables the use of renewable energy as distributed generation and encourages customers to reduce their carbon fuel–based power consumption.

The major requirements of smart grid communication [1] are as follows.

3.3.1   Quality of services (QoS)

Latency, which is the time delay between the cause and the effect of some physical change in the system being observed, bandwidth, and response are three main components of QoS. Smart grid communication technology is characterized by real-time operation, and monitoring and/or metering data should reach the control center within a very short time (response and latency), in milliseconds, and the bandwidth requirements are increasing due to more message transfers.

3.3.2   Interoperability

Interoperability can be defined as diverse systems working together, exchanging information using compatible parts. This phenomenon enables two-way communication and coordination between different equipment of the smart grid. The National Institute for Standards and Technology (NIST) was the first to focus on interoperability with the formation of the Smart Grid Interoperability Panel (SGIP) in 2009. It develops protocols and standards for the coordination and interoperability of different smart grid devices and components.

3.3.3   Scalability

Scalability in a smart grid communication network can be obtained by using Internet Protocol (IP)-based networks. Smart grid communication requires the inclusion of many devices and services and also real-time operation and monitoring of energy meters.

3.3.4   Security

The communication infrastructure of a smart grid is vulnerable to security attacks as the equipment is interconnected. Security concerns include attacks from disgruntled employees, industries, or terrorists, and security threats due to human error, equipment failure, and natural calamity. If a system is vulnerable to attacks, it allows the attacker to penetrate through the communication networks, gain access to the software, and change the settings to destabilize the grid.

3.3.5   Standardization

A smart grid uses many standards in the field of generation, transmission, distribution, customer, control, and communication. IEEE has defined these standards and issued guidelines for the use of the new technologies by smart grid. IEEE P2030 is the standard group created by IEEE, and this group mainly focuses on three fields: power engineering technology, communication, and information technology. The power engineering group focuses on standards for interoperability in smart grid; the information technology group works on the security, privacy, data integrity, interfaces, and so on; and the communication group works on the communication requirement between devices.

3.4   SCADA communication topologies

The devices in a SCADA system communicate with each other to operate the system effectively, and the devices are connected to one another in many ways, depending on the requirements. The topologies used for SCADA communications can be defined in two ways: physical, how the wires are physically connected, and logical, how the information is transmitted through the network.

3.4.1   Point to point and multi-drop

Two devices can be physically connected in two ways, the first of which is point to point where a dedicated communication link connects the two devices. The whole capacity of the link is used by the two devices for communicating. In multi-drop (multi-point), a single communication link is shared by more than two devices. The channel is shared by all the connected devices in two ways. In time sharing, specific time slots are allotted for each device. In spatial sharing, the devices use the channel simultaneously by sharing the channel capacity. Figure 3.2 presents point-to-point and multipoint links.

   When two or more links are used to connect devices (nodes), they form a network topology based on the way in which the devices are connected geometrically. The commonly used topologies are bus, ring, star, and mesh or a combination of these. With the advent of smart grid and larger systems, networks like LAN, WAN are also in use in power systems.

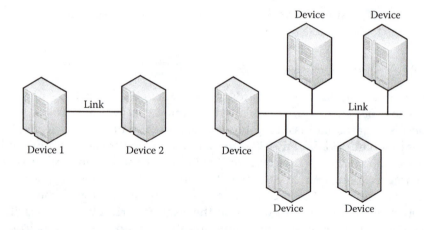

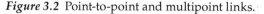

Figure 3.2 Point-to-point and multipoint links.

3.4.2 Bus topology

Bus topology is flexible, commonly used for the master station communication, and can cater to any communication technique, master slave, peer to peer, and so on. Each node is connected to a single or redundant bus that carries the message, the nodes pick up the messages that are intended for each individual node, and if any message is left without being accepted by any node, it is terminated electrically at the end of the bus. Bus topology is reliable, and any node failure will not affect the communication in the bus, and at the same time, the number of nodes can be increased or decreased easily. Node-to-node communication is possible, and this topology is not dependent on the master. Bus topology has some inherent disadvantages, as the failure of the bus is difficult to pinpoint, messages not picked up by a node are lost at the end as they are not returned. The bus may be busy during heavy traffic conditions, and a node may not be able to send messages on time.

3.4.3 Ring topology

In ring topology, all the nodes including the master form a ring, or closed loop, and the messages are transmitted from node to node in one direction. The message, if not accepted by any node, returns to the sender which suffices as an acknowledgement. Direct node-to-node communication is feasible in this topology, and any node can be a master. The major disadvantage is that the failure of any one node disrupts the whole network. Increasing and reducing the number of nodes is also a problem, as communication has to be stopped, and fault detection and isolation are also difficult.

3.4.4   Star topology

The star topology has a master that is the central hub, connected to the nodes by links. This is an easy configuration to develop, maintain, monitor, and trouble shoot. Adding and removing nodes is easy; however, this does not support direct communication between the nodes. The major disadvantage is the fact that in case of the failure of the master station, the entire network fails.

3.4.5   Mesh topology

Mesh topology is also used which is an improvement over the ring; redundant links make the network more reliable. Partially connected and fully connected mesh are used, depending on the redundancy level required.

3.4.6   Data flow: Simplex and duplex

The data flow between two devices can happen in two ways. In simplex flow, the data flow is only one way, and one device can transmit data to the other device, but the second device can only receive. In duplex flow, both devices can communicate over the link, which again can be done in two ways: half duplex and full duplex. In half duplex, both devices can communicate and receive, but not at the same time. A device can start communication using the whole channel, and the other device will receive and will start communication after the channel is free. In full duplex, both devices can transmit and receive at the same time. This can be done by having two dedicated channels between the devices, one for sending and another for receiving. Full duplex can also be achieved by dividing the channel capacity in two by suitable methods.

3.5   SCADA data communication techniques

3.5.1   Master-slave

In the master-slave mode of communication, one device acts as the master that controls the communication and the timing. All other devices can communicate only if the master initiates and allows the communication. Slaves cannot communicate with each other independently and can communicate only when permitted by the master. This technique can be used on any topology, and priorities are assigned for collecting data in some systems. This system uses the communication resources at a minimum, as the master has to initiate it which slows the speed of communication. A SCADA master will initiate communication from the slave remote terminal units (RTUs) and intelligent electronic devices (IEDs).

3.5.2   Peer-to-peer

In the peer-to-peer mode, when an event occurs, any device can initiate communication with any other device in the network, and all devices are equal, although sometimes a bus administrator is used to control traffic. When used in SCADA systems, the SCADA master station will still receive the majority of the data and initiate control commands; however, other devices will also have the capability to start communication. Network communication can still happen even if the master fails. Star topology does not support peer-to-peer, as all the connections terminate at the master and inter-node communication is not feasible. Peer-to-peer uses the communication resources in a better fashion; however, when the number of nodes rises, the performance decreases.

3.5.3   Multi-peer (broadcast and multicast)

The multi-peer technique allows communication of an active device with other devices in the group in two ways: broadcast and multicast. In broadcast, an active device sends a message to all other stations, master and slaves included, which is unacknowledged. In multicast, an active station sends messages to a group of devices, which are predefined, and the message is unacknowledged.

3.6   Data communication

As discussed in Chapter 1, the signals from the field, both analog and digital, are acquired by the sensors or transducers and reach the RTU/IED, and the analog signals get converted to digital by the analog-to-digital converters. The communication system of the SCADA has to transfer this binary data to appropriate monitoring centers, whether substations or state or national utility control centers.

This section discusses data communication which is the exchange of data between two devices, one device at the remote power system equipment or component to be monitored or controlled, and the other at the control center or substation, via some form of transmission medium. A detailed discussion of the transmission medium is presented in Section 3.10.

3.6.1   Components of a data communication system

1. *Message*: The message is the information (data) to be communicated, which could be values, switch positions, numbers, pictures, sound, video, or a combination of these.
2. *Sender*: The one who sends the message—the RTU/IED/substation computer, telephone, video camera, and so forth.
3. *Receiver*: The one to whom the message is destined, the front-end processor (FEP)/communication front end (CFE) of the master station, substation, and so on.
4. *Medium*: The physical path by which a message travels from sender to receiver (e.g., twisted pair wire, coaxial cable, fiber-optic cable, microwave, radio wave, etc.).
5. *Protocol*: A protocol is a set of rules and conventions that govern data communications and represents an agreement between the communicating devices. Two devices may be physically connected, but for data communication between the two, the devices should agree upon or understand the same protocol. A protocol defines what is communicated, how it is communicated, and when it is communicated.

3.6.2   Transmission of digital signals

The digital data from a device are to be communicated to another device via some physical medium, and for this purpose, the digital data are first converted to a digital signal for transmission. This process is called *encoding*. Line coding is a technique used for converting the digital data into a digital signal and at the receiving end, the signal is decoded to retrieve the digital data. The digital signal is generally nonperiodic, as the digital data are not in any specific pattern of zeros and ones.

The transmission of a digital signal is done in two ways: the first method is to transmit the digital signal directly, which is referred to as *baseband communication*, which requires a low pass channel with wide bandwidth. The other way of transmitting a digital signal is to convert it to an analog signal. In the frequency domain, a periodic digital signal, which is rare, will have infinite bandwidth and discrete frequencies, whereas a nonperiodic digital signal will have infinite bandwidth and continuous frequencies. A digital signal is a composite analog signal with frequencies varying from zero and infinity. The resulting analog signal, which represents the digital signal, can be transmitted using broadband communication.

3.6.2.1   Baseband communication

Baseband communication is referred to as direct transmission of the digital bit stream. This method is generally used with transmission over copper circuits for short distance and for optical fiber communication. The method implemented is called *on-off keying*. In this technique a 1 is transmitted when a voltage or current signal is applied on the communication media and 0 when no signal is applied. The effectiveness of this method can be measured by the receiver's ability to decode the signal or reconstruct it. This technique when used with copper circuits gives lower rate distance product. Thus, to compensate, this distortion of the received signal, either the distance between transmitter and receiver is kept small or a lower data rate should be used for transmission. Repeaters or the equalization filters (matched with the characteristics of communication media) could also be used to eliminate this distortion in the signal.

3.6.2.2   Broadband communication

Broadband communication changes the digital signal to a composite analog signal for transmission, by modulation technique. A sine wave is used as a carrier to transmit the digital signal. For a sine wave, three specific attributes, frequency, amplitude, and phase, can be defined, and by changing any one of the attributes, a different wave is created. The digital signal, which carries the data, is used to change frequency, amplitude, phase, or a combination of amplitude and phase of an electrical signal, and these are the mechanisms used in broadband transmission. When the frequency is changed, the phase is *frequency shift key* (FSK); for amplitude variation, it is *amplitude shift key* (ASK); and for phase shifting by the digital signal, it is *phase shift key* (PSK). However, the most popular technique in use is the quadrature amplitude modulation (QAM) where the phase and amplitude of the analog signal are varied appropriately by the digital signal.

3.6.3 Modes of digital data communication

The transmission of binary signal, which represents the binary data, can be accomplished by parallel mode or serial mode:

- Parallel mode: multiple bits are sent with each clock pulse using multiple parallel channels
- Serial mode: one bit is sent with each clock pulse, which again is subdivided into
  - Synchronous transmission
  - Asynchronous transmission

3.6.3.1   Synchronous data transmission

In synchronous data transmission, a clock signal is transmitted along with the transmitted data via a separate wire. This clock signal can also be included with the modulation technique. The data bit timing is identified by the transmitted clock signal. There is no added start or stop bit with each byte. In synchronous data transmission the bit stream is combined in a "frame." This frame contains multiple bytes without any gap between each byte. In this form of data transmission a SYNC character identifies the start of the transmission, which is placed at the start of the transmitted message.

USART (Universal Synchronous/Asynchronous Receiver Transmitter) is a decoding device that searches for the SYNC character and receives the data bytes. The receiver receives the bit strings and separates the strings into bytes or characters for decoding the message.

The synchronous data transmission is faster than asynchronous transmission as no extra bits are added in the message byte, and thus fewer bits move across the media.

3.6.3.2 *Asynchronous data transmission*

Asynchronous transmission follows a specific agreed-upon pattern and is not accompanied by a clock pulse, as in synchronous transmission. The message is sent in streams of usually a byte (8 bits) along with a start bit and one or more stop bits. The start bit, which is usually a zero, alerts the receiver about the arrival, and the stop bit (s) which is generally a one, ends the message. There could be gaps in between message streams, when the link may be idle; however, the arrival of the next stream will again start with a zero.

A device UART (Universal Asynchronous Receiver Transmitter) encodes and decodes the serial data. The UART at the sending end acquires one or more bytes from the processor and puts them in shift register, adds the start bit, error check bits, stop bit, and processes to transmit the data over communication media. The transmission generally starts from the least significant bit (LSB).

UART at the receiving end acquires the start bit on which it sets the time for the rest of the receiving bits. Asynchronous here means transmitter and receiver pairs are not synchronized at the bytes level but are synchronized within each byte at the bit level. As UART receives the start bit, it is independent of the start bit of any other byte. Thus a total of 10 bits are used to transmit 8 bits of message (except error check bits). In other words, 20% of bandwidth of communication media is utilized for timing purpose in asynchronous mode.

3.6.4 *Error detection techniques*

In data transmission some errors due to distortion in the signals can occur owing to loose connections, noise, lightning, and so on. Some techniques are used to detect these errors, and the most commonly used techniques in power system SCADA are parity check, checksum error detection, and cyclic redundancy check (CRC) error detection.

3.6.4.1   Parity check

This is the simplest technique used for error detection. In parity check a single bit is added to the data transmission path. Two types of parity check could be defined here: even parity and odd parity. This parity type is user selectable. With *even parity*, parity is selected such that the number of 1s of the message and the parity bit are added to an even number. For example if a user has selected even parity and the numbers of 1s in the message are odd, then the parity bit would be 1. And if the number of 1s of the message are even, then the parity bit would be set to 0. With *odd parity*, parity is selected such that the number of 1s of the transmitted message and the parity bit is an odd number. For setting the value of odd parity we follow the similar phenomenon as explained above for even parity. The error detected by parity check is communicated to the processor by UART or USART.

3.6.4.2   Checksum error detection

Checksum error detection is a common technique used to check the errors in a message. The "checksum" of the message bytes which are to be sent is computed. Checksum is the numeric summation of all the bytes of a message. The sender adds this checksum to the end of the message and transmits it to the receiver. The receiver evaluates the checksum again on receiving the message, and the sender-side checksum and receiver-side checksum are then compared. The receiver accepts the data if the values match, but otherwise rejects it or requests that the data be re-sent. The carries occurring on adding the bytes should be ignored. Checksum is not an efficient solution for error check, but the errors are minimized to a greater extent.

3.6.4.3   Cyclic redundancy check (CRC)

Cyclic redundancy check is an error detection technique widely used in local area networks (LANs) and wide area networks (WANs). CRC is so named because it has both linear and cyclic properties. In CRC, a reminder is added to the data blocks to be transmitted, based on the polynomial division of the data block content. A CRC encoder at the sending end computes a fixed-length binary sequence called the *checkvalue* for each block of data, which is added to the data to form a cordword. The cordword, which contains the data and the checkvalue, is transmitted to the receiver. At the receiving end, the cordword is read, and CRC is performed on the data block by the CRC decoder and compared with the checkvalue, and if in agreement, the data are accepted.

3.6.5 Media access control (MAC) techniques

Communicating devices in a network access the network and send or receive data as per a set plan or are governed by a set of rules. These rules constitute the media access control (MAC). The MAC layer in a communication protocol defines these rules and is a sublayer of the data link layer in a protocol, as explained in Section 3.7.1. However, MAC is discussed separately, as it is essential for a power automation engineer to understand these types of media access. The following types of techniques are used for media access:

Selection Techniques: The device to transmit the data is selected in a convenient manner or sequence, which includes schemes like polling and token pass.

Reservation Techniques: Here the same channel is split into subchannels using physical mechanisms like frequency division multiplexing or time division multiplexing, and a virtual channel is reserved for each node for data transmission.

Contention Techniques: In the contention schemes, the channel is used by the node which is alert and gets hold of the channel for transmission, and schemes include Aloha and CSMA/CD.

The following short descriptions present some of the schemes used in power system SCADA protocols.

3.6.5.1 Polling

In a master-slave configuration, the master requests data from a slave in a specific preset sequence and receives the data in a sequence. Priorities can be set for some slaves, which can be polled faster if the slave node has critical data. If the slave does not respond, the request is re-sent, and after a number of requests, as set by the program, the slave is marked as nonresponding. Polling can be initiated on any topology and physical media. However, it is an inefficient way of media usage for obvious reasons. The major advantage is that collision seldom happens and failure of the channel is detected instantly. The programming is easy and the system is reliable. However, a major handicap is that the slaves cannot communicate with each other, as data must go to the master and then be redirected, which makes it slow, and also interrupts from a slave cannot be handled.

3.6.5.2 Polling by exception

In this case, the event that has changed after the last polling is requested by the master, and hence with no event change happening, no data are returned. Switch positions change once in a while during a fault or some other disturbance, and most of the time, the positions are unchanged.

This is a more efficient way of communication. In this case, there can be a provision for time tagging of the event, indicating the exact time of the change, which is very important for power system performance evaluations. For analog points, only when the value of the point changes beyond its significant dead band between scans is the point reported.

3.6.5.3   Token passing

In the token passing technique, the nodes are numbered in a sequence order so that a logical ring is formed and a token, usually a message, passes around the ring. If a node or station has something to report, it will access the token, and the specific station has complete access to the channel for transmission of data. Once the transmission by the station is complete, the token is released and any other station that has to report can access the token, and the process goes on. Token pass can be used on both ring and bus topologies, where in a physical ring, the token passes around and any break in the channel causes disruption. In a bus topology, the token passes in a logical ring among the stations, and the loss of one node will not affect the communication between other nodes.

This scheme is much more reliable and faster than the polling scheme, as peer-to-peer communication is possible and there will be no collisions. This scheme is usually used by systems where the data transfer requirements may vary from node to node. The disadvantage is that sometimes a station with an urgent message will have to wait until the token is available to it, and communication failure detection is slow. To detect communication failure, techniques such as background polling or integrity polling are used by the traffic administrator, where the nodes are polled continuously at a slower rate to check the health and data status.

3.6.5.4   Time division multiplex media access

In this MAC, each station will get a fixed time slot for data transfer (time-division multiplexing [TDM]) and deterministic time responses are possible without collisions. This is used in bus or ring topology and a traffic controller controls the traffic in the system. The major disadvantage is that interrupts cannot be handled by this method, and it is difficult to detect a communication failure where background polling will have to be done.

3.6.5.5   Carrier sense multiple access with
collision detection (CSMA/CD)

In this scheme, each node checks the channel (multiple access) whether it is free and cannot start the transmission until the channel is free (carrier sense). However, due to the propagation delay in the physical medium, more than one node may transmit messages and there will be collisions. The transmission is stopped by the node when a collision is detected (collision detection), and the message is retransmitted after a random period. In this technique the channel is used most efficiently; however, each node should have the ability to detect collisions and avoid collisions and message retransmission and recovery schemes. CSMA/CD is used by Ethernet communications (IEEE 802.3 standard). The major advantage is that peer-to-peer communication is possible and variation in data transfer requirements can be handled well by this scheme. A centralized traffic controller is not necessary and priority access for urgent messages can be built into the protocols. The major drawback is the data collision and detection, avoidance, and recovery schemes have to be well developed at each node. Communication link failure is difficult to detect, and nondeterministic times of data transfer are further drawbacks.

3.7   SCADA communication protocol architecture [3,4,5]

Communication protocol defines the format in which the data is transferred from one device to the other, by which the communication becomes easier, both devices can decode the data received, and multiple data can be transferred using the same channel by the techniques discussed earlier.

As discussed, the power system data, say a voltage value, now a series of zeroes and ones after the digitization, constitute *bits*, and a series of 8 bits is a *byte*. A series of bytes constitutes a *data frame*, which includes the voltage data to be transmitted. In a traditional master-slave communication, the data are sent in a traditional message frame with a start/stop bit, control function, and error detection code added, as shown in Figure 3.3. This is the case when the dedicated point-to-point communication link is available. When the bus topology becomes more complex or when the data are to be sent over a network, which is generally the case nowadays, a lot more information such as the source and destination address, time stamping, control, and advanced error detection and correction information have to be added to the raw data. Adding such information to the data requires compliance with some specific rules; otherwise the receiving device will have difficulty in understanding and decoding the real data, and the result can be disastrous especially for a SCADA system with online monitoring and control functions to be implemented.

Every protocol defines a set of rules by which the data which are to be communicated are enveloped with source and destination addresses and error-checking mechanisms, which are sent over traditional communication channels. These use the traditional protocols, as will be clear from later discussions. If the data transfer is using a WAN, additional network and application layers will have to be added for successful communication. Examples of traditional protocols are Modbus, IEC 60870, and modern protocols include IEC 61850 and DNP3 (IEEE 1815) to some extent.

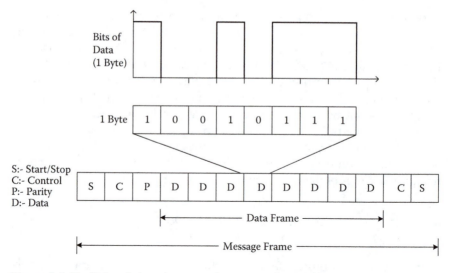

Figure 3.3 Traditional data frame and message frame for communication of a digital signal.

A protocol defines the exact process by which the additional layers are added to the data, and this process varies from protocol to protocol. This led to a large number of protocols evolving as per specific user requirements in the beginning. However, interoperability was compromised as proprietary protocols flourished and international organizations stepped in to define common formats for protocols.

The functionality of each layer of the communication protocol is defined clearly in the Open System Interconnection (OSI) reference model of 1984 issued by the International Organization for Standardization [ISO] and modified in 1994. Transmission Control Protocol/Internet Protocol (TCP/IP) also developed simultaneously, and some of the SCADA protocols are based on TCP/IP. The OSI model was adopted for SCADA communication as Enhanced Performance Architecture (EPA) by the International Electrotechnical Commission (IEC) and open SCADA communication protocols developed using EPA. In recent times, IEC 61850 was developed with full OSI layers for power system SCADA applications.

The following sections discuss the OSI model, the EPA and TCP/IP models, and later the discussion focuses on specific protocols.

3.7.1　OSI seven-layer model [5]

OSI is the open system interconnection that is responsible for network communication. An open system can be defined as a set of protocols that enables communication between two different systems without any

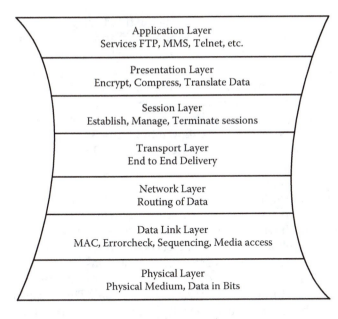

Figure 3.4 OSI model with layers and functionalities.

changes in the system's internal logic. OSI is a layered hierarchical architecture and has seven separate but functionally related layers. Each layer defines a part of the process for moving information from one system to another. The seven layers are physical (layer 1), data link (layer 2), network (layer 3), transport (layer 4), session (layer 5), presentation (layer 6), and application (layer 7). During the communication between two machines, a particular layer of one machine communicates with the same layer of another machine by means of protocols. The processes of that particular layer which communicates are called peer-to-peer processes. In a single machine, a layer uses the services of the layer just below it and provides the services for the layer just above it in the OSI model, as shown in Figure 3.4.

In the process of communication at the higher layer, the message moves down through the layers at the sending device and reaches the receiving device and then moves up through the layers. During the data transmission at the sending device, each layer receives the data from the layer above and adds its own information and transmits it to the layer below. At the receiving end the layers receive the message and extract the data meant for it and transfer the rest of the data to the above layer. Thus, the data are unwrapped layer by layer at the receiving device.

These seven layers can be grouped in three: the first group constitutes the *network support layers* which include physical, data link, and network layers. The *user support layers* include the session, presentation, and application layers. The *transport layer* is the third category. The OSI model represents an hourglass with expanded top and bottom and a narrow middle. This implies that the same data can be transported using multiple physical layers like twisted pair, microwave, or fiber-optic cable, and can have multiple application layers. However, all applications have to agree on a common set of networking protocols defined by the middle layers of network, transport, and session, for using the network, which can run over multiple physical layers, as shown in Figure 3.4.

The functions of each of the seven layers of OSI model are described here:

1. *Physical Layer*: The physical layer is concerned with a bit stream transmission over a physical medium. It defines the electrical characteristics of the interfaces between the devices, such as the frequency and the level of the optical or electrical pulses to be transmitted and also the type of transmission media, such as fiber optic or microwave. For the data to be transmitted, the bit stream data should be converted into electrical or optical signal. This encoding is also defined by the physical layer. It defines the data transmission rate, the synchronization of the bits at the sender and receiving end, the connection of the devices to the media (i.e., point-to-point or multipoint), physical topology (how devices are connected for making a network), i.e., mesh, ring, bus, star or hybrid topology. The transmission mode, simplex, half duplex, or full duplex, is also set by the physical layer.

2. *Data Link Layer*: The data link layer makes the frames of the bit streams. Frames are defined as the manageable data units of a bit stream. The job of this layer is to move the frames from one node to the next. The data link layer has the capability of physical addressing, which means if the frames are to be sent to different systems on a network, then a header is attached to the frames which defines the sender and/or receiver address. It has flow control, error control, and access control mechanisms. Flow control avoids the overflowing of the receiver with the data if the data rate is not the same for sender and receiver. For error control a trailer is added to the end of the frame, as discussed in the error control mechanisms in the previous section. It recognizes the damaged or lost frame and resends it. Access control is the control over the link by any particular device, if more devices are connected to the same link. The media access control (MAC) layer is a sublayer of the data link layer in the OSI model, and the way the media is accessed by the communicating device is decided by the MAC. It links the data link layer with the physical

medium, and the methods of media access used in SCADA systems are discussed in Section 3.6.5.

3. *Network Layer*: The network layer provides delivery of message packets from source host to destination host when the two systems are connected to different networks. If the communication is between devices that are on the same network, there is no need for a network layer. Logical addressing and routing are the main functions of the network layer. When the communicating devices are on different networks, to address the source and destination devices the network layer adds a header to the packet data. To connect these different networks, connecting devices are used which are the routers. Routers create internetworks and route the data packets to the destination host.

4. *Transport Layer*: This layer is responsible for process-to-process message delivery. The transport layer ensures the delivery of messages in order. This layer has some functions including service point addressing, segmentation and reassembly of message, and connection, flow, and error control.

   Service point addressing or port addressing delivers the message to the correct process among multiple processes running on a single computer. For this purpose, the transport layer adds a header to the message which defines the port address. For the reassembly of the messages at the receiving device, the message is divided into segments, and each segment will have a sequence number. On receiving the data packets at the receiving device, the packets are reassembled with the help of the sequence number and the lost packets in transmission are identified and replaced. The transport layer data transmission can be either connectionless or connection oriented. The flow control and error control mechanisms are provided so that the message reaches the receiver without any loss or damage. Retransmission is done for error correction.

5. *Session Layer*: The session layer provides the dialog control and synchronization between sender and receiver processes to establish a healthy communication. The communication mode could be one way at a time (half-duplex) or two ways at a time (full-duplex) as provided by session layer. Synchronization is another service of session layer to ensure the proper delivery of messages to the receiver. For this purpose a stream of data is divided into equal fixed-length messages and a checkpoint or synchronization point is added to each of these fixed-length messages. Thus it acknowledges the delivery of these messages independently. If a loss or damage occurs to a particular fixed-length message, then only that particular length message is retransmitted instead of the full-length message.

6. *Presentation Layer*: The presentation layer provides the services of data translation, encryption, and compression. In the communication process the message is converted into bit streams and then transmitted, and since different systems use different encoding methods for the bit streams, a presentation layer is required. It converts the encoded message obtained from the sender to a common format and then transmits it to a receiver. At the receiver side, the presentation layer converts the common format into a format that is understandable and used by the receiver. For some important and sensitive messages, data privacy is required, and *encryption* is the process that converts the transmitted message into a different format before the message is transmitted over the network. At the receiver's side the original data or information is fetched from this data, which is called *decryption*. The multimedia message transmission (i.e., text, audio, video) requires the data compression, which reduces the number of bits contained in a particular message.

7. *Application Layer*: It is the topmost layer, which provides the network access to the user, where services like file access and transfer, mail services, and directory services are provided. Network virtual terminal allows a user to log on to a remote host. When the user tries to log on to the host, the user computer communicates with the software virtual terminal which in turn communicates with the host and thus provides accessibility to the user. This application allows the user to read the remote host files and make changes, manage and control host files, and also sometimes fetch the host files for use in the local computer. Mail and directory services that provide global information about various services and objects are also part of the network layer.

The voltage data transport over the OSI model can be described to explain the seven-layer model. Figure 3.5 shows the construction of the message frame in device 1 and transmission via the physical medium and the reduction of the message frame in device 2 to retrieve the message.

The voltage, as discussed earlier, that is acquired from the field is converted to digital form of bytes in the RTU/IED. As the bits of this voltage data move down the OSI stack, each layer will add information to it at the sending end (device 1), while the same information is removed at the receiver end (device 2) to retrieve the voltage bytes. The application layer will add some meaning to the data, by specifying that it is a voltage message, is in kV, utilizes phase-to-phase voltage, and adds related information. The presentation layer then does the encoding of the data into a common format such as ASCII, which is easily understood by all devices, and at the receiver presentation layer, the ASCII format is decoded into the specific format understood by the receiver. Encryption, if required, is

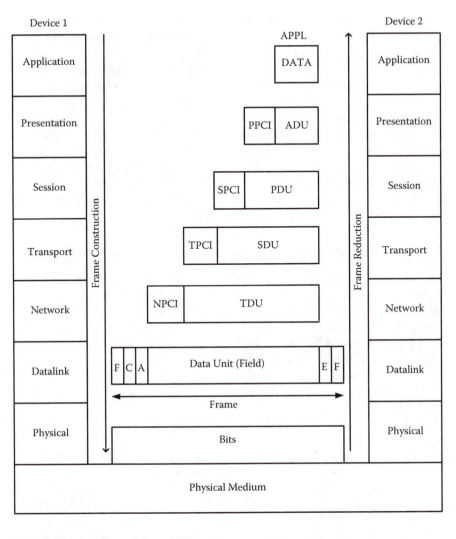

PPCI: Presentation Protocol Control Information   F: Flag
SPCI: Session Protocol Control Information   C: Control
TPCI: Transport Protocol Control Information   A: Address
NPCI: Network Protocol Control Information   E: Frame Check sequence

Figure 3.5 OSI message frame construction and reduction.

also done by this layer. The session layer will initiate the transmission of this data, which would have been broken down into fixed message sizes and numbered, by defining the start and end of the session at both the user ends, by ensuring that the message is received in the same sequence, so that the voltage value can be recreated. If not, the retransmission of the lost part is ensured. The transport layer will control the transmission by adding check points and synchronization points for proper end-to-end delivery, as different applications will be running on the same system, say different screens may be open at both ends. The network layer adds the information regarding the destination network, if the source device is on a different network. The data link layer creates the frames out of the bit steams received from the network layer. The physical addressing is done as a footer, and the frame sequence is established. This layer will also add error control, flow control, and access control mechanisms, as shown by the header and footer attached. The frames of voltage thus formed are converted into the appropriate physical state such as light pulse, or microwave signals for transmission over the corresponding physical medium such as fiber-optic cable or microwave, respectively.

3.7.2 Enhanced performance architecture (EPA) model

For SCADA systems and IED communication, it is clear that all seven layers of the OSI model are not necessary all the time. For this purpose the IEC (International Electrotechnical Commission) has introduced a reduced form of the OSI seven-layer model, which is a three-layer model called the Enhanced Performance Architecture (EPA). The three layers are physical, data link, and application layers. Physical and data link layers are called hardware layers, and the application is the software layer.

EPA, when used over a network or networks, adds a pseudo-transport layer to assist network communication.

Functions of the EPA model layers are as follows:

1. *Physical Layer*: The physical layer is the bottom layer of the EPA model just as in OSI model; it is the physical media for transmission of data bits. The physical layer converts each frame into a bit stream to be sent over the physical media and keeps check of transmission of one bit at a time. The physical layer provides the services of connection between sender and receiver, disconnection, sending message and receiving message. The connection topology could be point-to-point, multi-drop, hierarchical, or with multiple masters. The data communication procedure through these topologies can be either half duplex, where the communication is one way at a time, or full duplex, where the communication is two way, with two transmitters and receivers.

2. *Data Link Layer*: While the physical layer is concerned with the transmission of a single bit of data, the data link layer deals with groups of data and their reliable transmission. This group of data is called a *message frame*, as discussed earlier. For security and reliability purposes, an acknowledgment is also sent for the receipt of data. Data

link also provides the services of data flow control and error check. For error detection and correction, error checking codes are introduced in the data.

3. *Pseudo-Transport Layer*: The pseudo-transport layer does the combined function of network and transport layer of the OSI model. Network function is concerned with the routing and data flow over the network from sender to receiver. Transport function includes proper delivery of the message from sender to receiver, message sequencing, and error correction. This function of the transport layer is limited when compared to the OSI layer, and that is the reason why it is called a pseudo-transport layer. The message structure of the transport header includes a start bit and a stop bit which identify the sequence of frames and a six-bit sequence counter.

4. *Application Layer*: The application layer serves the end user directly and will add meaning to the data received from the process. It helps the user to do functions like file transfer and network access. Figure 3.6 gives the EPA model along with the OSI model for comparison.

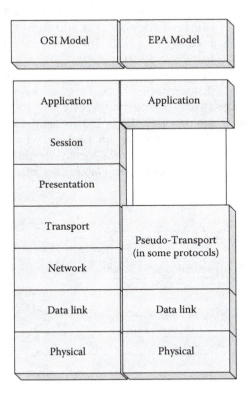

Figure 3.6 EPA and OSI models.

3.7.3 TCP/IP model

TCP/IP is a four-layer model, the layers being host-to-network, Internet, transport, and application. The functionality of TCP/IP layers can be compared with the OSI layers functions. The combination of physical and data link layers of OSI is similar to the host-to-network layer of TCP/IP. The Internet layer is doing the same job as the network layer in OSI, as shown in Figure 3.7. The fourth layer of TCP/IP is the application layer which is similar to the combination of session, presentation, and application layers of OSI, and the transport layer which is performing the tasks of the session layer of OSI.

TCP/IP is a hierarchical protocol that implies that the lower-level protocols support the upper-level protocols, and these protocols can be used according to the system requirement by mixing and matching. These protocols are not interdependent as in the case of the OSI model, where each layer performs specific tasks and provides or uses service only to or from the adjacent layer above or below.

A brief description of these layers is as follows:

1. *Host-to-Network (Physical and Data Link Layers)*: At this layer TCP/IP supports almost all network interfaces and transmission media.
2. *Internet (Network) Layer*: This is the internetworking layer of TCP/IP which supports internetworking protocol (IP). IP uses four protocols: Address Resolution Protocol (ARP), Reverse Address Resolution Protocol (RARP), Internet Control Message Protocol (ICMP), and Internet Group Message Protocol (IGMP).

   Internetworking protocol (IP): IP is the transmission mechanism of the TCP/IP. IP transmit data in packets, which are called *datagrams*. These packets are transported separately, and hence IP is a connectionless protocol. Datagrams reach the destination

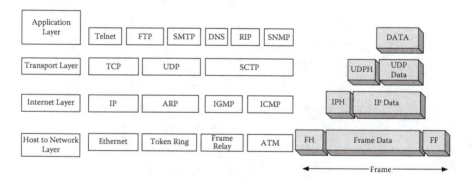

Figure 3.7 TCP/IP layers and the message frame.

through different routes, and there is a possibility of being out of sequence and duplicated. IP does not have the facility for error checking or routes tracking, making it an unreliable data transmission service. It is a kind of protocol that provides transmission but with no guarantees.

Address Resolution Protocol (ARP): In a LAN-type physical network, each device has a physical address that is mentioned on the network interface card (NIC). ARP finds the node's physical address when the network address is known.

Reverse Address Resolution Protocol (RARP): As the name implies, it is the reverse of ARP and is used to find out the host's Internet address when a computer is connected to the network for the first time and its physical address is known.

Internet Control Message Protocol (ICMP): If there is any problem in the received datagram, then ICMP is used to send the error reporting messages and query back to the sender.

Internet Group Message Protocol (IGMP): IGMP is used to send a message to more than one recipient (group of recipients) simultaneously.

3. *Transport Layer:* The transport layer protocols enable the transmission of data from one running process to another process, and it is called *process-to-process* delivery. The major transport layer protocols are TCP (Transmission Control Protocol) and UDP (User Datagram Protocol).

TCP: TCP is the most commonly used transport protocol. It provides connection-oriented, reliable data transmission. Connection oriented implies that proper connection must be set up between the two devices, before data transmission. During data transmission operation, TCP divides the whole data stream into segments. These data segments are assigned with sequence numbers for reordering of the data if required, on receipt at the receiving end, and with acknowledgment numbers for acknowledging the receipt. Thus, TCP provides reliable data transmission along with duplicate data suppression, flow control, and congestion control.

UDP: User Datagram Protocol is a simple but unreliable protocol. It transports the message with much less interaction between sender and receiver in comparison to TCP or Stream Control Transmission Protocol (SCTP). UDP is a connectionless service, where the datagrams sent are independently and are not numbered, which implies that the datagrams can follow different paths to reach the destination even when they are coming from the same source device and going to the same destination. UDP does not have flow control and the receiver can overflow with messages; hence, this is not suitable for transmission of bulk data. UDP uses a checksum error control mechanism.

    SCTP: Stream Control Transmission Protocol (SCTP) is basically designed for internet application. SCTP is a reliable and message oriented transport layer protocol.

4. *Application Layer:* This layer is the combination of session, presentation, and application layers of the OSI model. It supports various protocols like File Transfer Protocol (FTP), Domain Name System (DNS), Routing Information Protocol (RIP), Telnet, Simple Mail Transfer Protocol (SMTP), and so forth. Figure 3.7 presents the TCP/IP model.

3.8   Evolution of SCADA communication protocols

The early SCADA systems which were all analog, used direct current pulses for signaling, and analog information was sent using current loops. When digital communications started, analog values were also digitized and sent over channels with parity checks for error detection. Later CRC was introduced; however, each manufacturer used to define its own communication protocol, and in the early 1980s around 100 SCADA protocols were competing with each other. The confusion created was tremendous as the proprietary protocols were hardware-specific and posed integration problems during expansion of equipment and systems. The value of formulating standard protocols and open systems was clearly understood by the manufacturers and utilities. A number of user groups and organizations started working toward standardization in the early 1980s, and by 1986 standard SCADA protocols began to be introduced. It may be noted that the bases for these protocols were the proprietary vendor-specific systems which slowly became industry practices and were so popular that they went on to become industry standards. Modbus is an example, which was a proprietary protocol introduced by Modicon in 1979 for programmable logic controllers. It became so popular that it went on to become an industry standard by fact (de facto). By law (de jure), standards are developed by national or international standards development organizations such as American National Standards Institute (ANSI), National Institute of Standards and Technology (NIST), Institute of Electrical and Electronic Engineers (IEEE), International Electro-technical Commission (IEC), and so on. As examples, the development of IEC 60870 and DNP (Distributed Network Protocol) occurred during the same time frame (late 1980s to early 1990s) as open SCADA protocols.

    The major attraction was that devices from different vendors could communicate with each other, making the systems vendor independent. The added advantages of reduced software costs, removal of protocol convertors, shorter delivery periods, and the like, accelerated these developments. Thus, the market evolved with fewer widely accepted, standard, open protocols, which provided many long-term benefits like easy system expansion, which saved money, faster adoption to new technologies, and product life expectancy improvement.

The Institute of Electrical and Electronic Engineers (IEEE) and the International Electro-technical Commission (IEC) started working on SCADA communication protocols. The Electric Power Research Institute (EPRI) published the Utility Communication Architecture (UCA) report in 1991.

3.9   SCADA and smart grid protocols [6,7,8,11–20]

As discussed earlier, SCADA systems in the early 1980s had a variety of proprietary protocols that streamlined into a few de facto and de jure standards from the early 1990s, and the process of refining some of the protocols is still going on. The following sections discuss a few of the popular protocols in use today in the power system SCADA scenario. ICCP (IEC 670-6) is the international standard for one control center to talk with another control center. For communication from the master station to the field equipment, DNP3 (IEEE 1815) is used in North America and the IEC 870-5-101(T101) serial and 104 (TCP/IP) are used in Europe and by European vendors. For communication between field equipment IEC 61850, DNP3 (IEEE1815) and Modbus are used.

3.9.1   Modbus

Modbus started as a protocol for communication with the PLC and went on to become the most widely accepted protocol for industrial electronic devices. It is a de facto standard openly published, and approximately 40% of communication within industrial appliances uses the Modbus protocol. In SCADA systems, Modbus is used for communication between master stations and remote terminal units.

Modbus uses the layers 1, 2, and 7 of the OSI model and is based on polling for media access control. The error detection uses a cyclic redundancy check. Modbus uses a master-slave technique and only a master device can initiate the transaction or query. The messages are transmitted in frames, and the frame format consists of four fields: address, control, message, and error check. Figure 3.8 illustrates a Modbus application data unit.

3.9.1.1   Modbus message frame

A query or request by the master will have the address of the slave device in the address field and uses one byte. The address range allowed by the Modbus is 1 to 247. For the broadcast message to all devices it uses address 0. The slave places its own address in the response field so that the master can know which slave is responding. The function field will have the control task to be performed by the slave device, such as read or write one byte or read event counters. The data field will vary in length, and the master

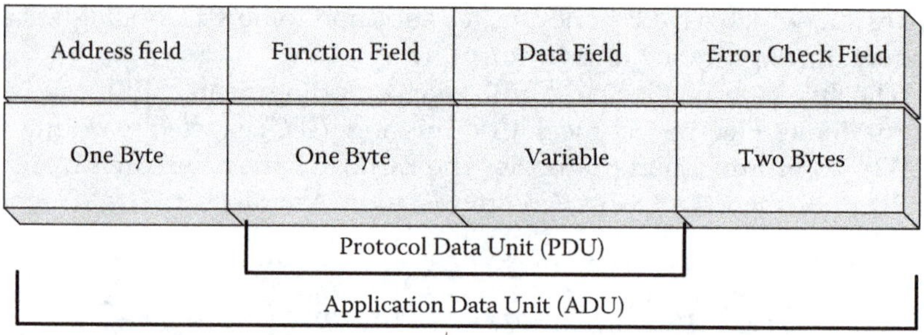

Address field	Function Field	Data Field	Error Check Field
One Byte	One Byte	Variable	Two Bytes

Protocol Data Unit (PDU)

Application Data Unit (ADU)

Figure 3.8 Modbus application data unit (ADU).

request will have the information required to complete the function. The error check field will have the CRC code, and the slave checks the data for error before accepting and executing the command.

The response frame sent by the slave will have similar fields and send information as requested by the master. In the communication, the master can address individual slaves or can send a message to all the slaves at one time. The slaves respond to the master when addressed individually, as a query expecting a response, whereas no response is sent when a broadcast message is sent to all the slaves. The number of slaves a master can handle is up to 247, but typically a master will have a few slaves in a SCADA system. Some characteristics like frame format as discussed above, the frame sequence, error checking (CRC), and exceptions are fixed for the Modbus protocol, whereas the transmission mode, transmission media, and characteristics are selectable.

There are two asynchronous transmission modes in a Modbus network: ASCII and RTU. RTU mode is compact and faster and is used for normal operations. The message frame discussed above is used by the RTU mode. American Standard Code for Information Interchange (ASCII) mode is used for testing of the system. The message frame is seven characters long and the address field is two characters.

Modbus can also be implemented using TCP/IP using Ethernet, which is common now for station LANs. Modbus Plus is a proprietary protocol by Modicon.

3.9.2   IEC 60870-5-101/103/104

IEC 60870-5 was introduced for SCADA telemetry by the IEC Technical Committee 57. This is an open protocol, which is applicable to telecontrol equipment of the SCADA system basically for industry level. Initially the use of IEC 60870-5 protocol was started in the European countries. The structure of this standard is hierarchical and has six parts, each part having different sections, and it has four companion standards. Main parts of the standard define the fields of application, whereas the companion standards elaborate the information regarding the application field by giving specific details:

IEC 60870 Telecontrol equipment and systems Part 5
    Sections of Part 5:
        5-1 Transmission protocols
        5-2 Link Transmission Procedures
        5-3 Structure of Application Data
        5-4 Definition of Application Information Elements
        5-5 Basic Application Functions
    Companion Standard of Part 5:
        5-101 Basic Telecontrol Tasks: 1995
        5-102 Transmission of Integrated Totals: 1996
        5-103 Protection Equipment: 1997
        5-104 Network Access: 2000

These companion standards may be referred to as T-101, T-102, T-103, and T-104, where *T* stands for telecontrol.

T-101 and T-103 use master-slave commination for multi-drop or bus topology, poll data by cyclic polling technique using the data link layer (MAC), and use parity check as well as checksum error detection techniques. The data error is decreased by these checks in IEC 60870 protocols. The specialty of these protocols is that they are developed for specific use in the power industry, and the application functions for T-101 include station initiation, data acquisition, cyclic data transmission, clock synchronization, parameter loading, and so forth, for a remote substation. T-103 is specifically for protection functions and handles all such functions as status indications of circuit breakers, type of fault, trip signals, auto recloser, relay pickup, and so on. T-104 is the networked version for use in networked circumstances.

3.9.2.1   Protocol architecture

IEC 60870-5 protocol is based on the Enhanced Performance Architecture (EPA) model described earlier. The EPA model has three layers: physical, data link, and application. A user layer is added to the top of the EPA model to provide interoperability between equipment in a telecontrol system. This four-layer model is used for T-101 and T-103 companion standards. For companion standard T-104, which is the network adaptation, some additional layers are included from the OSI model. These are network and transport layers that are essential for the networked architecture. This networked architecture is useful for the transportation of data and messages over the network. Thus, a non-networked version of the model is used for T-101, T-103, and the networked version for T-104, as

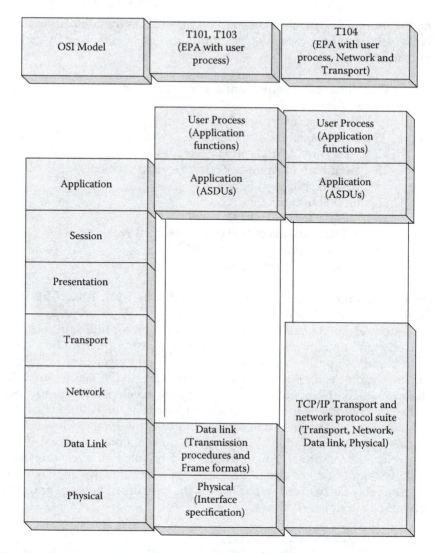

Figure 3.9 IEC 60870 protocol layers and the OSI model comparison.

shown in Figure 3.9. It may be noted that the lower four layers of T-104 are now the TCP/IP suite for networking applications.

3.9.2.2 IEC 60870 message structure

The message frame of T-101, which follows FT 1.2 frame format specified by IEC 60870-5-1, has three options: variable length, fixed length, and a message with a single control character.

The data received from the application process is the application data unit (ASDU), which contains the information objects format as given in

		START FRAME	LPDU
Start S		LPCI	(Link
Length L		(Link Protocol Control	Protocol
Length L		Information)	data Unit)
Start S			
Link control C			
Link address A			
Link address A			
Data unit identifier		ASDU	
Info Object Identifier	Information object 1	(Application layer data service Unit)	
Information elements			
Time tag			
Information object 2			

Information object n			
Checksum CS		STOP FRAME LPCI	
End E			

Figure 3.10 FT 1.2 ASDU, LPCI and LPDU structure.

Figure 3.10. The data link layer adds the link protocol control information (LPCI) to this data as header and footer from the link protocol data unit (LPDU). As can be seen, the message frame uses checksum error detection. The figure shows the variable length message frame structure LPDU where each block represents one character, the LPCI header with a start, length (message length), link control, link address (twice), and user data (ASDU) which can be up to 253 octets (8 bits) and the LPCI footer with checksum and end characters. Once the LPDUs are formed, the data are sent via the physical medium as octets.

Each octet is preceded by a star bit, a parity bit (even), and a stop bit for transmission. The data link layer prepares a series of such octets with start, parity, and stop bits for transmission for each LPDU. Hence, this protocol uses parity at the octet level and checksum at the message frame level for error checking. It may be noted that this protocol uses link addresses for the link data. A variable length frame carries up to 253 octets of link user data, and a fixed length frame carries up to 5 to 6 octets, which are used for only control commands and without any user data.

These protocols are used for operating the control equipment in a substation and fetching to transmit the information to the master devices in a large number of substations across the world.

3.9.3 Distributed network protocol 3 (DNP3)

DNP3, Distributed Network Protocol version 3.3, is a telecommunication open protocol initially developed by Westronics in Calgary, Alberta, Canada. DNP3 is also based on the EPA architecture and uses the frame format FT3 specified by IEC 60870-5. The lower layers of physical and data link defining the communication between devices are similar to IEC 60870-5-101 and the higher levels of data units and functionality are different. DNP3 uses cyclic redundancy check for error detection. It has larger data frames and can carry larger RTU messages.

3.9.3.1 DNP3 protocol structure

The structure uses the basic three-layer EPA model with some added functionality. It adds an additional layer named the pseudo-transport layer. The pseudo-transport layer is a combination of network and transport layer of the OSI model and also includes some functions of the data link layer. Network function is concerned with the routing and data flow over the network from sender to receiver. Transport function includes proper delivery of the message from sender to receiver, message sequencing, and corresponding error correction. This function of transport layer is limited when compared to the OSI layer, and hence the name *pseudo-transport layer*, as shown in Figure 3.11.

3.9.3.2 DNP3 message structure

A DNP3 message starts with the information from the user data, which could be an alarm, values of variables, control signals, program files or any other data. DNP3 does not impose any restriction on the size of data to be transferred. The data are broken down into smaller manageable sizes called application size data units (ASDUs). The application header called application protocol control information (APCI) of 2/4 bytes in length is added to the ASDU, which makes an application protocol data unit (APDU). The APDU is referred to as a transport service data unit (TSDU) within the pseudo-transport layer. As mentioned earlier, since the ASDU in DNP3 could be large in size, due to the large user data, this is broken down into transport protocol data units (TPDUs) that have up to 250 bytes of data to fit into the data link frame. The data link layer adds a fixed header, which is 10 bytes, to the user data as shown in Figure 3.12, and the CRCs, which complete the FT3 frame, which is termed the link protocol data unit (LPDU) for transfer through the physical medium in octets.

DNP3 supports some system topologies such as peer-to-peer, multiple master, multiple slave, and hierarchical with intermediate data concentrator. DNP3 is used in many industries like electricity, oil and gas, security, and water, while IEC 60870-5 is limited to the electrical distribution industry.

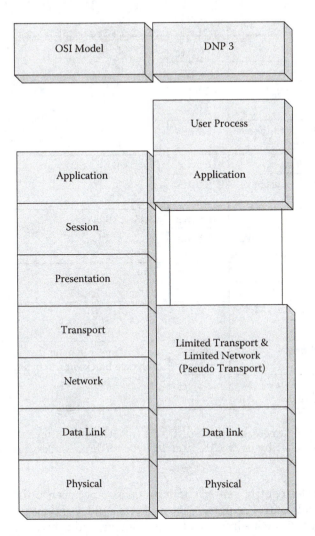

Figure 3.11 DNP3 and OSI model layers.

DNP3 is popular in the Americas, Australia, and parts of Asia and Africa, whereas IEC 60870-5 is popular in Europe and parts of Asia and Africa.

3.9.4 Inter-control center protocol (ICCP)

ICCP is IEC 60870 part 6, and it defines the communication specifications for sending telecontrol messages between two stations on a WAN. With the SCADA systems expanding and now controlling the gigantic transmission systems across the world, ICCP is the standard protocol that is

Start	2 bytes	Block 0, Fixed length header
Length	1 byte	10 bytes
Control	1 byte	
Destination	2 bytes	
Source	2 bytes	
CRC (for the above header)	2 bytes	
User data	16 bytes	Block 1, 18 bytes
CRC for the above user data	2 bytes	
User data	16 bytes	Block 2, 18 bytes
CRC for the above user data	2 bytes	
-------------------		-----------------------

-------------------		---------------------

User data	Up to 16 bytes	Block n, up to 18 bytes (n up to 16) 16th data block with 10 bytes, a total
CRC for above user data	2 bytes	of 250 bytes of data

Figure 3.12 FP3 message frame of DNP3 (LPDU with up to 250 bytes of data (TPDU), up to 32 bytes of CRC and 10 bytes of link header for a total up to 292 bytes).

used for inter-control center communication, between utilities, regional and national control centers, ISOs, and large independent power producers. ICCP carries real-time data for monitoring and control of large power pools. The standard development started in 1991 by the IEC Technical Committee 57 (Working Group 3) and TASE 1 (Tele-control Application and Service Element) was released in 1992 and later TASE 2 with MMS (Manufacturing Message Specifications) which is widely used now, was released as IEC 60870-6-503 (2002-04) also known as Inter-Control Center Protocol (ICCP).

Some of the relevant sections of the protocol are

- IEC 60870-6-2 Use of basic standards (OSI layers 1–4)
- IEC 60870-6-501 TASE.1 Service definitions
- IEC 60870-6-502 TASE.1 Protocol definitions
- IEC 60870-6-503 TASE.2 Services and protocol

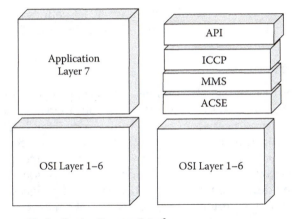

API : Application Program Interface
ACSE : Application Control Service Element

Figure 3.13 ICCP structure.

- IEC 60870-6-504 TASE.1 User conventions
- IEC 60870-6-601 Functional profile for providing the connection-oriented transport service in an end system connected via permanent access to a packet switched data network
- IEC 60870-6-602 TASE transport profiles
- IEC 60870-6-802 TASE.2 Object models

ICCP is a client-server protocol, and any control center can act as both client and server. The communication channel can be point to point or over a network. The clients can establish multiple connections with the same server at different levels, so that real-time priority data can be transferred faster than nonpriority data.

ICCP uses layer 7 of the application layer of the OSI model with MMS for messages. ICCP specifies the control center object formats and methods for data request and reporting, whereas MMS specifies the naming, listing, and addressing of variables and message control and interpretation mechanisms, as shown in Figure 3.13.

3.9.5   Ethernet

Ethernet standard evolved from ALOHA network which was set up to link islands in Hawaii. In 1980, the Ethernet consortium released the Ethernet blue book 1, and in 1983, IEEE released the IEEE 802-3 standard which is based on carrier sense multiple access with collision detection (CSMA/CD) on a LAN-based network. Ethernet standard uses only the physical and data link layers of the OSI models with CSMA/CD for medium access control. The IEEE standard 802-3 defines a large number of cable types which are used for networking using this standard.

3.9.6   IEC 61850

As seen from the previous discussions, standard protocols exist at different levels of the power system for different applications. In the substation, for instance, communication to a higher hierarchy is by ICCP, the downstream communication is by DNP3, IEC 60870 series, and so on. Hence, in 1995, the IEC working groups started work on a protocol for complete substation automation, which could be used for all applications and functionalities, and the developed protocol is IEC 61850. The objectives included the design of a single protocol for a complete substation considering modelling of different data required and to define basic services required for future proof mapping of data. Highest interoperability with vendor independent systems and devices is a major priority of the protocol, and it also defines a common method to store and format the complete data. Testing standards for compliant devices are also defined by the protocol. The data models specified in IEC 61850 can be mapped to many other protocols, the current mapping is to manufacturing message specifications (MMSs), generic object-oriented substation events (GOOSEs), sampled measured values (SMVs), and at a later stage, to Web services.

The standard is organized into 10 parts: IEC 61850-1 gives introduction and overview; part 2 gives the glossary; part 3 specifies the general requirements; part 4 gives the system and project management; part 5 states the communication requirements for functions and device models; part 6 provides the configuration language for communication in electrical substations related to IEDs; part 7 details the basic communication structure for substation and feeder equipment and has subparts to define the principles and models, common data classes, compatible logical node classes and data classes, and the communication networks and systems for power utility automation; part 8 provides mapping to MMS; part 9 covers specific communication service mapping (SCSM); and part 10 gives details of the conformance testing of compliant devices.

For substation automation implementation, the information exchange between the process level devices and substation is necessary. This is possible only when all the devices are compatible with the protocol. Various protocols are introduced for SCADA telecontrol operations but none of them fully incorporate the IED interoperability when the IEDs are from different vendors. IEC 61850 advances the substation automation functionality to a new level where IED interoperability is key.

The main function of SAS can be broken down in subfunctions, which are defined as the logical nodes, and the logical nodes reside in logical

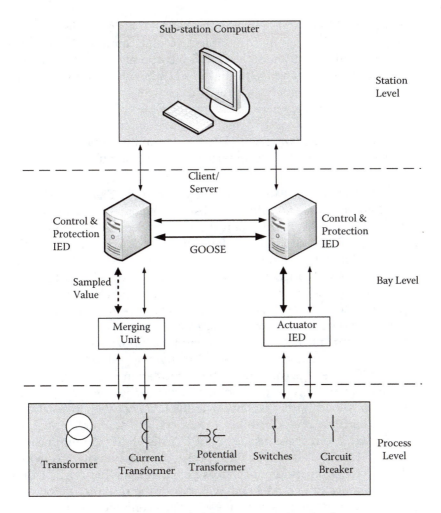

Figure 3.14 IEC 61850 GOOSE and client server communication.

devices like IEDs. One or more subfunctions can be performed by one IED. These subfunctions can be distributed within three levels: process, bay, and station levels, as shown in Figure 3.14. Process-level functions deal with switchgear equipment, including extraction of data from the field through sensor or transducer and transmission to the upper bay level. They also incorporate the function of sending the control function from bay level to the switches or field devices. The bay level includes four main functions in SCADA (i.e., metering, monitoring, protection, and control). These applications can be performed by a single device, the IED installed at the bay level. Bay-level equipment collects the data from its own bay or from different bays and performs the required action on its own primary equipment. Human-machine interface and communication function are included in the station level and incorporate communication to the remote control center and to the bay-level IEDs. At the station level, the process-level data are analyzed, and the control command is sent to the process-level equipment.

The main features of IEC 61850 include interoperability, object-oriented data model, switched Ethernet communication technology, GOOSE messaging, and SCL (standardized configuration language).

1. *Object-Oriented Data Model*: This object-oriented data model specifies the data model for different substation automation functions and devices. Logical nodes contain the data, and data have some data attributes. The object data model of IEC 61850 defines a set of rules to create more logical nodes and common data classes, thus extending the protocol functionality out of the substation, viz., distribution automation, feeder automation, and communication to control center.
2. *GOOSE*: GOOSE messaging provides peer-to-peer communication. Client-server communication is necessary for managing the process-level equipment. The critical commands of a power system such as interlocking or trip command are directly mapped to the link layer. A GOOSE message contains all the information regarding the state of an event and is repeated until the subscriber receives it. A GOOSE message allows faster response over the LAN network in a station.
3. *SCL*: The functional capability of devices and the interconnection of devices are represented using SCL. This allows communication interoperability. These SCL files specify the functions to the devices. The devices from different vendors exchange data through SCL files. The standard data model and SCL reduce the reconfiguration cost. Thus, extension of the system is easier and economical for increasing smart grid applications.

IEC 61850 supports both client-server and peer-to-peer communication, as shown in Figure 3.14. IEC 61850 is in use now for communication at the bay level extensively by utilities across the world, with expansion to process level and station level occurring rapidly.

3.9.7 *IEEE C37.118: Synchrophasor standard [9,10]*

This protocol is basically introduced for the synchronized phasor and frequency measurement in power systems. It also gives the methods for the verification of the measured values and time stamping facility.

Initially it was introduced in 1995 with the name of IEEE standard 1344, which had some shortcomings; hence, an improved version was introduced in 2006 as C37.118 2005. This protocol was for steady-state measurements, and the latest version is C37.118 2011 introduced in 2011, which has two parts: C37.118.1-2011 and C37.118.2™-2011, IEEE standard for synchrophasor measurements for power systems and IEEE standard for synchrophasor data transfer for power system, respectively. The C37.118.1-2011 protocol is basically introduced for the synchronized phasor and frequency measurement in power systems in different substations. It also gives the methods for verification of the measured values. C37.118.2-2011 defines the data exchange mechanism between phasor measurement unit and phasor data concentrator, and specifies the messaging format for real-time application.

3.9.7.1 Measurement time tag from synchrophasor

Time tagging is done using the measurement time, with respect to Coordinated Universal Time (UTC), which used to be Greenwich Mean Time (GMT) until 1972. The time tagging includes three numbers, namely: SOC (second-of-century), FRACSEC (fraction-of-second) count, and message time quality flag. The SOC counts from midnight (00:00:00) of January 1, 1970, to the current second. To keep the system synchronized with the UTC, the leap seconds can be deleted or added accordingly.

3.9.7.2 Reporting rates

The PMU supports data reporting rates which are submultiples of power line frequency (i.e., 10 frames/s, 25 frames/s, or 50 frames/s at 50 Hz frequency, and 10, 12, 15, 20, 30, or 60 frames/s at 60 Hz frequency). These data rates are selected by users. The communication media may vary; normally optical fiber cables are used.

3.9.7.3 Message structure

The first transmitted frame is SYNC which is of 2 bytes and identifies the starting of the frame, which is needed for synchronization purposes. The second frame is FRAMESIZE (2 bytes) followed by IDCODE (2 bytes) which provides the information about the source data, SOC (4 bytes), FRACSEC (4 bytes), and the last transmitted word is CHK. CHK is used for message verification (i.e., it is errorless and not corrupted). The message structure is shown in Figure 3.15. The data frame has phasor values, frequency, ROCOF (rate of change of frequency), analog values (voltages and currents), and digital values (status points). Here cyclic redundancy codes (CRCs) are used for error check, which forms the last frame of 2 bytes.

No.	Field	Size(Bytes)	Comment
1	SYNC	2	Sync byte/frame type/version number
2	FRAMESIZE	2	Byte in frame
3	IDCODE	2	ID Number
4	SOC	4	Time Stamp
5	FRACSEC	4	Fraction of Second and Time Quality
6	STAT	2	Bit-mapped Flag
7	PHASORS	4 x PHNMR Or 8 x PHNMR	Phasor estimates 1 φ or 3φ voltage and current; +ve, -ve, zero seq
8	FREQ	2/4	Frequency (fixed or floating point)
9	DFREQ	2/4	ROCOF (fixed or floating point)
10	ANALOG	2 x ANNMR Or 4 x ANNMR	Analog data
11	DIGITAL	2 x DGNMR	Digital data
	Repeat 6-11		Fields 6-11 are repeated for as many PMUs as in NUM_PMU field in configuration frame
12+	CHK	2	CRCCCITT

Figure 3.15 IEEE C37.118 message frame.

3.9.8 Wireless technologies for home automation [12,13,14]

There are some technologies available for WHANs (wireless home area networks):

1. ZigBee
2. Wi-Fi
3. Z-wave
4. Insteon
5. Wavenis

With the increasing popularity of ZigBee and Wi-Fi, the following sections discuss some aspects of these technologies.

3.9.8.1 ZigBee

Home automation is introduced at an early stage to empower the consumer to manage electricity consumption. With the advent of wireless technologies, home automation is becoming more popular. Consumers can control lighting and home appliances and manage power consumption economically. ZigBee is a wireless standard based on IEEE 802.15.4 and is a radio frequency (RF) communication protocol. It can operate in the frequency band of 868 MHz, 915 MHz, and 2.4 GHz, and the data rates are 20, 40, and 250 kb/s, as defined by the IEEE standard 802.15.4. The standard defines the physical and MAC layer, and ZigBee, developed by ZigBee Alliance, defines the layers above. This is a wireless technology for short-range and low-data rate and provides services for home and commercial building automation, consumer electronic devices, and energy management.

There are mainly two application profiles of ZigBee standard:

1. ZigBee home automation profile
2. ZigBee smart energy profile

The first application profile is in the area of residential and commercial mainly for light control, security, and home appliance control. The second profile is for the demand response and consumer load control. This requires the communication facility available between the ZigBee smart energy wireless home automation network (WHAN) and the communication network of the power supply company. The ZigBee smart energy profile requires more security features as compared to the ZigBee home automation profile.

3.9.8.2 ZigBee devices

ZigBee incorporates three kinds of devices: coordinator, router, and end devices.

1. *ZigBee Coordinator*: It creates the network and controls it for home automation. All home electronic devices are connected to the network through the coordinator. There are mainly two kinds of data transfer: one in which the coordinator receives the data sent by the devices, and other in which the coordinator sends the data received by the devices. A third kind of data transfer also exists which is peer-to-peer in which data is transferred between two devices on a network.
2. *ZigBee Router*: It provides an alternate path for the data transfer in case of network congestion or failure of devices. It connects to the coordinator or to the other router so that the data transfer will be uninterruptable and the network area coverage will be extended.
3. *ZigBee End Devices*: These are controlled electronic devices. They send or receive the data to or from the coordinator. They are connected to the coordinator or the router for the data transfer. They do not perform any routing function.

ZigBee is a mesh-type network in which a device automatically discovers the nearest available path to transfer the data (a type of router). This kind of network is self-healing. Thus, the path of the communication will be zig-zag, which is similar to the path by which the bees communicate or send messages to the other bees in the hive, and hence the name *ZigBee*.

3.9.8.3 Wi-Fi

Wi-Fi uses IEEE 802.11 standard, and this technology makes it possible for a set of devices to talk to each other on a wireless LAN within a short distance. Any Wi-Fi–compliant device can connect to the network using a wireless network access point, the limitation being the distance covered. The distance is limited to 35 m indoors and 100 m outdoors, which can be expanded to kilometers by overlapping access points. Wi-Fi alliance is the consortium of companies that brings out Wi-Fi hotspot devices. Wi-Fi is used extensively in offices, buildings, and for home automation at a carrier wave of specific frequency in the range of 2.4 GHz ultra-high frequency (UHF) or 5 GHz SHF radio waves. The carrier wave from the transmitter carries the data packets, in Ethernet frames and will be received by all stations within the range. This technology is adapted by many automation platforms, as a variety of devices can communicate using Wi-Fi.

3.9.9 Protocols in the power system: Deployed and evolving

As discussed earlier, currently there are well-established communication protocols, de facto or de jure, at all levels of automation in power systems. Figure 3.16 attempts to explain the popular protocols and their domains.

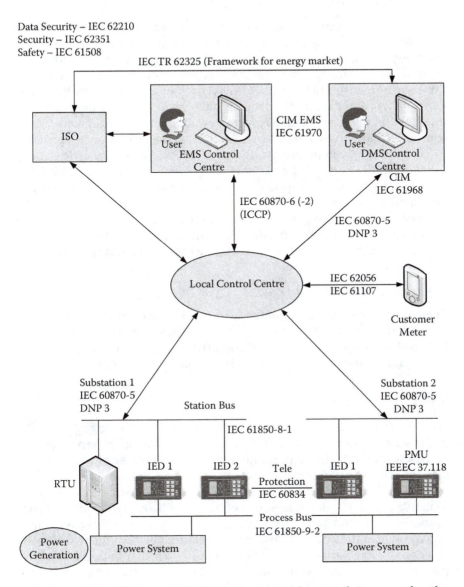

Figure 3.16 (See color insert.) SCADA and smart grid protocols in use and under development.

3.10   Media for SCADA and smart grid communication [2,1]

Once the communication requirements for automating the power system constituent, whether generation, transmission, distribution, or customer, are determined, the medium to be used for communication at different levels of automation has to be decided. The developments in communication have immensely helped the SCADA communication, as new technologies, options, and data transfer speeds have developed over the years. The Internet and the associated hardware development have also helped data communication to reach new levels of speed and accuracy at cheaper cost.

A variety of media are available for the system designer to choose from, depending on speed and data transfer requirements. The following section gives a brief description of the media used for SCADA and smart grid communications.

The media discussion is divided into two parts: guided, where there is a physical medium to carry the data, and unguided, where there is no physical medium to carry the data and the transmission is wireless.

3.11   Guided media [2,4,5]

There is a wide selection of guided media available for SCADA communication depending on the distance, terrain, and the money the utility wants to spend for a communication infrastructure. The following are some of the guided media used by SCADA and a smart grid system.

3.11.1   Twisted pair

Twisted-pair telecommunication cable has been in use for many years. As utilities have existing pole infrastructures, installation of aerial cable is more appropriate. The configuration contains a number of pairs of conductors and uses metallic (copper) conductors, each with its own plastic insulation that accepts and transmits signals in the form of electric current. In a pair of twisted cables, one of the wires is used to send the signal to the receiver, and the other is considered only as a ground reference. The receiver at the other end uses the difference between the two signals.

The construction of the twisted cable is such that both wires of the pair are equally affected by interference (noise) or cross talk which creates unwanted signals as the two wires are twisted and are not in parallel. If one wire is closer to the noise source in one twist, in the next twist, the other wire will be closer to the noise source; thus, both wires are equally affected by the noise. The advantage is that at the receiver side, the unwanted signals are mostly canceled out because the receiver calculates the difference between the two. In parallel wire construction, the effect of unwanted signals is different in both the wires because the distance of two wires from the noise or cross talk source is different (one is nearer and the other is farther).

There are two types of twisted pair cables: unshielded twisted pair (UTP) and shielded twisted pair (STP). In STP cable, a metal foil or braided mesh cover encases each pair of insulated conductors, and it prevents the penetration of noise or cross talk, making it bulky and more expensive as compared to UTP.

Twisted-pair cables are basically used in telephone lines to provide voice and data channels. Unshielded twisted-pair cables are commonly used. LANs also use twisted-pair cables.

Twisted pair is a cheap medium for short distances and can handle up to 1.54 MHz channel capacity. But this medium has some disadvantages as it is subject to breakage and water ingress. In twisted-pair cable, failures may be difficult to pinpoint, and it is also subject to ground potential rise due to power faults and lightning.

3.11.2 Coaxial (coax) metallic cable

These media also transport signals in the form of electric current. They can transmit high-frequency signals of several MHz, as the two conductors are constructed differently. The coax has a central core conductor of solid or stranded wire (usually copper) instead of having two wires. This core conductor is enclosed in an insulating sheath, which in turn is enclosed in an outer conductor of metal foil. This outer conductor works as a shield against noise and as a second conductor as well which completes the circuit. The outer conductor is encased in an insulating sheath, and the whole cable is protected by a plastic cover. Coaxial cable has much higher bandwidth, but the attenuation is also high in this cable as compared to the twisted-pair cable. That means the signal weakens rapidly, and to compensate for this we have to use the repeaters frequently.

Coaxial cable was used in analog telephone networks and later in digital telephone networks, but a large part of coaxial cable in the telephone network has been replaced with fiber-optic cables. The traditional television network also used coaxial cable for its entire network but later, most of the cables were replaced by fiber-optic cable. In today's hybrid network, coaxial cable is used in the consumer premises. Coaxial cables are also used in traditional Ethernet LANs.

The methods of installation of coaxial cable are underground, overhead, direct burial, and under build on an existing power line structure. Coaxial metallic cable has limited bandwidth requirement, saturated frequency bands for point-to-point radio, right of way available for its route, lower cost, more immunity to RF noise interference, and hence is economical for short distances. These media usually support voice, data, frame relay, switched T1, switched multimedia data services (SMDSs), fractional T1, and interoffice trunking. The disadvantages of coaxial cable are media breakage and water seepage and failures that may be difficult to pinpoint.

3.11.3   Optical fiber

A fiber-optic cable is made of glass or plastic and accepts and transports signals in the form of light. This medium offers high bandwidth and immunity from electromagnetic interference. The physical characteristics of the fiber depend on the propagation modes of light. Current technology supports single mode and multimode. Demand for single-mode fiber is increasing as compared to multimode because multimode has limited distance and bandwidth characteristics, whereas single mode supports higher signaling speed due to the smaller diameter.

The physical structure of the cable has a core member in the center and additional outer layers of cladding for providing support and protection to the cable against physical damage and effects of the elements over a long period of time. This technology has advanced tremendously, and now the commercially available fibers have losses less than 0.3 db/km. Designers consider fiber-optic technology for systems of 140 km or more without any repeaters because of low loss, as well as availability of suitable lasers and optical detectors. It supports communication services like voice, protective relaying, telemetering, EMS/SCADA, video conferencing, high-speed data, and telephone switched tie trunks.

There are three types of fiber cables available for use in SCADA and smart grid applications. One is optical power ground wire (OPGW) that has an optical fiber core within the ground or shield wire suspended above the transmission lines. All-dielectric self-supporting (ADSS) is another type of cable that has a long span of all-dielectric cable designed to be fastened to high-voltage transmission line towers underneath the power conductors. A third type of cable is wrapped optical cable (WOC) that is usually wrapped around the phase conductor or existing ground/earth wire of the transmission or distribution line. Aerial fiber-optic cable can be fastened to the distribution poles under the power lines.

Fiber-optic cable has wide bandwidth which is cost effective; thus, it is often found in backbone networks. For a hybrid network of coaxial and fiber-optic cable, optical fiber provides the backbone structure while coaxial cable provides the connection to the user premises. Some fast Ethernet LANs also use fiber-optic cable.

There are some major advantages of optical fiber cable: it has low operating cost and high channel capacity, and there are no licensing requirements. This cable is immune to electromagnetic interference and ground potential rise. It is light weight as compared to copper cable, and it has great immunity to tapping. The media has some disadvantages, as it is a novel technology and new skills must be learned, it requires expensive test equipment, and the cable may be subjected to breakage and water seepage problems. Optical fiber has unidirectional light propagation, and for bidirectional communication two fibers must be used.

3.11.4 Power line carrier communication (PLCC)

Power line carrier communication (PLCC) occurs when a power line that carries 50 Hz voltage and current is used to carry data signals also at a different frequency. This is an economical means of transmitting data over the existing power line, as additional media are not required. There are different PLCC techniques for different uses: as a power line carrier (PLC), distribution line carrier (DLC), and broadband over power lines (BPL).

3.11.4.1 Power line carrier (PLC)

PLC is the first reliable communication medium available to utilities. It uses the power feeder lines as communication media. PLC transmits the radio frequency signals in the range of 30 to 500 kHz. The main components of PLC links are transmitter and receiver terminals, coaxial cable, impedance matching devices, and coupling capacitor for insulation and to inject high-frequency signal onto the distribution line. Line traps are also installed on the power conductor to block the signals entering the substation through an undesired path. PLC equipment is located within the substation, and thus the security is very high. This medium supports services such as voice, telemetry, SCADA, and relaying communication on 220/230 kV, 110/115 kV, or 66 kV interconnected power transmission network at an available data transmission rate up to 9600 baud. There are two types of PLC: analog and digital. Digital PLC requires more maintenance as compared to analog and it is not recommended for noisy power lines. But digital PLC can be increased from one to three channels within the same RF bandwidth. Digital PLC has the capacity for three to four channels (e.g., two voice and one high-speed data), whereas analog PLC has the capacity for two channels (e.g., one voice and one "speech plus" low-speed data). The main disadvantage is that it is not independent of the power line. The availability of fewer channels may be a disadvantage of PLC, and it is expensive on the per-channel basis.

3.11.4.2 Distribution line carrier (DLC)

As the name implies, DLC uses the distribution line in the voltage range of 11 kV/22 kV/33 kV for the transmission of carrier signal in the range of 5 to 150 kHz. DLC supports a one-way requirement for direct load control and a two-way requirement for distribution automation. However, DLC has not been successful in distribution lines, as it requires many feeder traps, and transformers and noise levels are quite high in low-voltage distribution lines. The impedances of the distribution lines are also high, and the carrier frequency is also limited.

3.11.4.3   Broadband over power lines (BPL)

Broadband over power lines (BPL) works by coupling RF energy to the existing electrical power lines inside homes. In BPL technology, the principles of radio, wireless networking, and modems are combined, and a mechanism has been created where one can plug in the computer using BPL modems into any electrical outlet at home to have instantaneous access to high-speed Internet, instead of wiring additional data cables.

BPL is based on existing power line communications (PLC) technology, and to achieve high bandwidth levels, BPL operates at frequencies higher than traditional power line communications, typically in the range of 2 to 80 MHz. Many power line devices use orthogonal frequency division multiplexing (OFDM) to extend Ethernet connections to other rooms in a home through the power wiring. Adaptive modulation used in OFDM helps it to cope with noisy channels such as the home electrical wiring. OFDM is superior to spread spectrum or narrowband for spectral efficiency, robustness against channel distortions, and the ability to adapt to channel changes. BPL can also serve as a technology for home automation and for data access from a smart meter.

3.11.5   Telephone-based systems

Many telephone-based system are in use by the utilities for power system communication, for SCADA, telemetry, and smart grid applications. The following are a few of these options.

3.11.5.1   Telephone lines: Dial-up and leased

Dial-up connection provided by the telephone company, which is the same as that provided at home, allows temporary access, whereas a leased line is always available for use by the utility. The leased line can be ordered to be dedicated with separate routing for the utility, which is much desired, as the line will be laid for the utility and any change in performance can be tracked easily. Otherwise, the line will be routed along with the other customer lines and can be changed without prior notice to the utility.

The main advantages of leased circuit are that no communication expertise is required, and it is adaptable to changing traffic patterns. But this circuit is vulnerable to security attacks as it can be easily tapped in an unobtrusive manner. Also, the leased circuit can be rerouted in the telephone switch by a malicious intruder, and dial-up circuits can easily be accessed just by dialing the phone numbers from a public telephone network. These circuits are also subject to electromagnetic interference. Thus, proper isolation and protection should be provided to the telephone circuit against these vulnerabilities.

3.11.5.2   ISDN (integrated services digital network)

An ISDN is designed to combine digital telephony and data transport services. It is a switched, end-to-end, WAN. ISDN is used by larger businesses for the networking of geographically dispersed sites. Two types of services are available: ISDN basic access (192 kbps) and ISDN primary access (1.544 Mbps). Broadband ISDN (B-ISDN) is the next generation of ISDN, with data rates of either 155.520 Mbps or 622.080 Mbps.

As ISDN is a wired service, it is subject to electromagnetic interference. It also faces some security issue like rerouting of private services by any invader breaking into the telephone company equipment.

3.11.5.3   Digital subscriber loop (DSL)

DSL is used for data transmission over a standard analog subscriber line. DSL is based on the same technology as ISDN, but it is an economical system. It transmits moderately high bandwidth to small offices and residences. There are several types of DSL:

- *ADSL (Asymmetric DSL):* This transmits data and voice over ordinary copper pairs between the customer and telephone company. It supports data rates ranging from 1.5 to 8 Mbps downstream and 16 to 640 kbps upstream. Filters are used to separate the digital and analog streams at the telephone company central office and customer premises. ADSL can operate up to 6000 m.
- *SDSL (Symmetric/Single-Line DSL):* This can carry T1 on a single pair. (T1 is a high-speed digital network of 1.544 Mbps developed by AT&T for pulse-code modulation [PCM] wire transmission.)
- *HDSL (High-Speed DSL):* This is in use for point-to-point T1 connection. It can carry T1 and FT1 in both directions.
- *VDSL (Very High–Speed DSL):* It has the highest speed implementation to date. It can support 52 Mbps data downstream over short ranges but can attain full speed only up to 300 m.

The main advantage of DSL is its wide availability and competitive pricing. The disadvantage associated with it is its service limitation over circuit length, and as a wired service it is also susceptible to electromagnetic interference.

3.12   Unguided (wireless) media

3.12.1   Satellite communication

Satellite systems offer high-speed data services. The satellites are positioned in geostationary orbits above the Earth's equator and thus have continuous coverage over a particular area of the Earth. These systems contain a number of radio transponders that receive and retransmit frequencies to ground stations within their "footprints" (coverage areas) on the Earth's surface. A ground network facility tracks and controls the satellites and an antenna situated at the ground station receives signals from the satellites.

Very small aperture terminal (VSAT) technology is advancing steadily, where a much smaller antenna, as small as 1 m, can be used for the communication. The key to successful communication through satellite is to place satellites in the appropriate orbit. These are geosynchronous Earth orbit (GEO), medium Earth orbit (MEO), and low Earth orbit (LEO). The GEO system is mostly in use.

The GEO system requires large parabolic antennas to keep satellite transponder power levels to a manageable level, because the satellite is distant. Hughes built the first geosynchronous orbit (GEO) communication satellite in the early 1960s. These satellites do not need costly tracking antenna as they appear to stand still in the sky. A GEO communication satellite operates in an Earth orbit 22,300 mi (35,900 km) above the ground.

MEO satellites operate in orbits typically 10,000 to 20,000 km above the Earth's surface. A third technology is LEO which operates at a much lower altitude of 500 to 2000 km. Due to the smaller distances involved, lower power levels are required. A small Earth station is required as compared to GEO.

The main advantages of satellite systems are that they cover a wide area, and access to a remote site is easy. The error rate in this medium is low, and as the communication network changes continuously, the satellite system is adaptable to these changes.

Dependency on a remote facility is a disadvantage of a satellite, as is the delay in transmission time. Leasing of the satellite communication channel is a permanent expense for a utility.

3.12.2   Radio (VHF, UHF, spread spectrum)

Radio waves are generally omnidirectional. The waves transmitted by an antenna are propagated in all directions, thus avoiding any alignment for the sending and receiving antennas. The sending antenna waves can be received by any receiving antenna. Radio wave is a good medium for long-distance broadcasting. There are different types of antennas based on wavelength, strength, and purpose of transmission, and there could be many receivers for one sender due to the omnidirectional characteristic (multicasting).

The very high frequency (VHF) radio band lies within the range of 30 to 300 MHz. This radio frequency is mostly used by mobile services. This communication system can be used for maintenance of vehicle dispatching systems in power system automation and also for SCADA/DMS, for which an exclusive frequency should be assigned.

The main advantage of VHF radio is that a particular frequency can be assigned for a particular service. Cost is lower than that for microwave radio, and it does not depend on common carrier and power lines. The disadvantages are the low data rate for digital communication, the limited transmission technique, and low transmission channel capacity.

Ultra-high frequency (UHF) radio communication has typically a frequency band of 300 to 3000 MHz. Generally, a 400 to 900 MHz frequency band range is considered for UHF radio. The varieties of UHF systems are point-to-point (PTP), point-to-multipoint (PTM), multiple address radio systems (MARS), trunked mobile radio, and spread spectrum.

A point-to-point UHF system is used for the communication between master stations and individual substations. PTM UHF system has a frequency band of 400/900 MHz. PTM (MARS) is a single-channel system that communicates with each of its remotes or slaves in sequence. SCADA and telemetry/data and voice (on a limited basis) are the services supported by PTM. Cost of this medium is less than that for PTP. In a MARS system, the data speed is less as compared to PTP because of the multipoint operation.

A trunked mobile radio system is generally used by electric utilities for voice communication (e.g., crew dispatch), but sometimes it is also used for data communication. The main services supported by this media are SCADA, distribution automation, mobile, and paging. This system has a communication controller that assigns an individual channel to the user when required. The frequency band used for this system is 400, 800, and 900 MHz. Installation of a 900 MHz multitrunk radio system with the controller or switch at the control center and base station located in a utility service area is in process.

Spread spectrum radio does not require any radio frequency license. In the US, 902–928 MHz band, 2.4 and 5.3 GHz band low power spread spectrum radios are in operation without any license. This medium supports distribution automation services effectively.

3.12.3   Microwaves

Microwaves are electromagnetic waves with a frequency range of 1 to 300 GHz. Microwaves support both the analog and digital transmission technology. This medium is unidirectional where antennas send out signals in one direction. Thus, the sending and receiving antennas should be aligned so the transmitted microwave can be narrowly focused.

The main advantage of microwaves is its unidirectional characteristic, as a pair of antennas can be aligned without interfering with another pair of aligned antennas. A broad microwave frequency range makes high data rate possible.

Microwave use has some disadvantages, too, as a certain portion of the frequency band requires special permission from authorities. Another disadvantage is that it requires clearance of line of sight. The towers with the mounted antennas should be in direct sight of each other. Sometimes repeaters are needed for long-distance communication.

Major usage of microwave systems is in the cellular phones services, satellite networks, and wireless LANs. They are basically in use where unicast (one-to-one) communication is required. Digital microwave use is costly for individual substation installations but can be considered as a high-performance medium for establishment of a backbone communication infrastructure.

3.12.4 Cell phone

There are various common carrier services available for digital cell phone operation. A conventional service is cellular digital packet data (CDPD). The existing analog cellular phone service works along with CDPD digital cell phone service, as the digital service has the same operating frequency as that of analog, which is 19.2 kbps. Cell phones were in service for many years but now new applications are emerging which use time division multiple axis (TDMA), code division multiple access (CDMA), global system for mobile communication (GSM), personal communication services (PCS), and so on. Some more advanced technologies are playing a crucial role in the advancement of wireless cell phone technology, the major one being wideband technology which includes EDGE, W-CDMA, CDMA2000, and W-TDMA. The pricing of digital cellular phone usage is higher as compared to analog service.

3.12.5 Paging

It is a low-cost technique that is appropriate for some utility applications. Its communication technique is basically for one-way outbound messaging but some also offers inbound services. It can also use the satellite channel to increase the service coverage area. A paging system uses data links, controller, and transmitter for efficient service. Sometimes a number of transmitters can be used to continuously increase the range of service coverage. Paging service is publicly accessible via Internet or dial-up modem. As a public network, the security of the paging application layer level is needed.

It is clear that for a SCADA or smart grid communication system designer, there is a wide variety of physical media available to choose from, depending on the application, data exchange rate requirements, amount of data, and future expansion plans.

3.13 Communication media: Utility owned versus leased

A variety of SCADA communication media, as discussed in the previous sections, have been deployed in the power automation scenario. This includes utility-owned media, which are constructed and managed by the power utility. The statistics of the utility owned versus leased media are as shown in Table 3.1.

Table 3.1

Wired		Wireless	
Utility owned	Leased	Utility owned	Leased
Power line	Leased telephone	Spread spectrum radio	Paging system
Dedicated line	TV cable	MAS radio	Cellular phone
Optical fiber			Satellite
Twisted pair			
Coaxial cable			

Technology	Utility-operated?	Application			
		To s/s	Within s/s	To feeder	To customer
Distribution power line	Y			X	X
Telephone wire	N	X			X
Dedicated metallic line	Y		X		
Optic fiber	Y	X	X		
Cable sheath	Y	X			
Pilot wire	Y	X			
Cable	N				X
Utility-operated licensed radio	Y	X		X	X
Unlicensed radio	Y	X		X	X
Public broadcast	N				X
PCS	N			X	X
Trunking radio	Y			X	X
Satellite	N	X			
Cellular telephone	N			X	X

Note: These are from John McDonald's course titled "Communication Issues."

Critical applications such as defense services communication and railway communication always use dedicated communication infrastructures built and maintained by the organization. This makes it easier to have complete control over the channel in terms of quality, safety, and reliability and provide enough bandwidth requirement at any time. Although it is easier to acquire leased lines than to design, erect, and maintain a communication channel, it will be worth the effort to own SCADA communication as it is a mission-critical application.

3.14 Security for SCADA and smart grid communication [15,4]

The LAN- and WAN-based digital control systems started emerging in recent decades in power system automation and have led to many security issues. Earlier systems were simpler master-slave configurations with dedicated communication systems, and data security issues were minimal. With enterprise-wide networking and data transfer, linking the automation systems with the corporate WAN has made utilities more vulnerable.

Data corruption due to noise and other transmission effects is dealt with by the built-in error-checking mechanisms in the communication protocol, as discussed earlier, with parity, checksum, and CRC being some methods. However, these measures built into the data link layer cannot provide protection against the inadvertent packet corruption due to the hardware to the channel used for communication.

Cyber security attacks that cause havoc could be in the form of authorization violation, eavesdropping, information leakage, interception or alteration of the messages, masquerading, replaying a message at inappropriate times, and denial of service by choking the channel.

The security threats to the system can be from external intruders and from inside the corporate network. External intruders can gain access to the system via the SCADA communication lines as well as the dial-up lines to some systems like IEDs. The lines can be privately owned or leased from a carrier and are vulnerable to eavesdropping and intrusions that can corrupt the data. The intruders could be members of general public or cyber criminals. Internal intrusion can happen inside the corporate WAN, as many unauthorized users can gain access to the data unless the utility is very strict with the password policy, privileges, and authentication measures. Employers of suppliers of substation equipment and software may gain access to critical data and can extract crucial information or cause damage unless prevented effectively from doing so.

Encryption is the method that can be effective to handle many of the attacks discussed, which is inbuilt in the OSI reference model, in layers of network or above. IP security (IPsec) protocol inserted at the IP level is an example. However, the real-time data transfer requirements and slow processors at the RTU/IED level limit the encryption. The processors may take longer to encrypt the message, and it may take longer to communicate longer, encrypted messages.

Cyber security discussions in power system SCADA are taking a little longer to turn around with positive results, but there is a need to accelerate the process, looking at the security threats in a traditional SCADA setup, as shown in Figure 3.17.

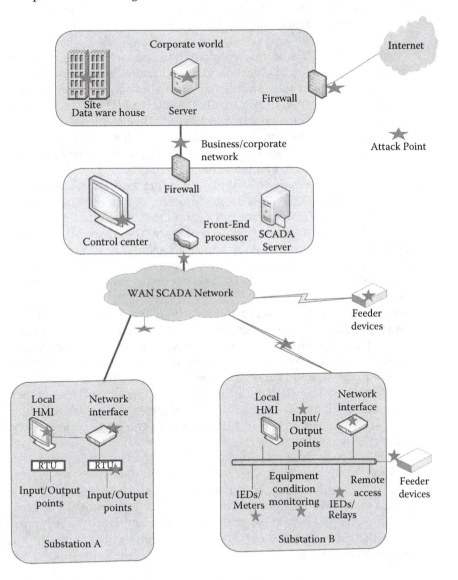

Figure 3.17 Security threat points in a SCADA network.

3.15   Challenges for SCADA and smart grid communication

SCADA and smart grid infrastructures are scattered over a wide geographical area and carry time-critical information to the substations and control centers. Thus, the communication systems pose serious challenges with regard to investment, bandwidth, speed of data transmission, reliability, and maintenance. Some of the serious challenges as the grid migrates to a smarter system are discussed here:

1. *Communication is Foundational*: Before there can be a smart grid there must be a strong grid. Two foundational technologies must be up to date and robust to support smart grid applications: communication infrastructure and information technology (IT) infrastructure. Communication is the enabling technology that enables automation. Key metrics for each smart grid application that the communication infrastructure must support with the required performance are response, bandwidth, and latency requirements.
2. *Efficiency of Communication*: With the smart grid envisaging bidirectional information flow from domestic devices to substations, from substation to local control centers and integrated control of transmission and generation in real time, the efficiency of communication is of utmost importance. To achieve this, better telemetry like PMUs with faster scanning, system-wide faster controls, enhanced computing facilities, and secure communication channels are required.
3. *Security*: As discussed earlier, the security of the communication systems is of utmost importance, and there is a need for stricter guidelines and procedures to ensure security. Specific security domains could be defined and maintained by the power industry, such as power plant domain, substation domain, telecom domain, real-time domain, corporate IT domain, and public, supplier, and maintainer domain. Security threats and vulnerability of the communication infrastructure should be addressed, and at the same time privacy should be maintained.
4. *Impact of Communication on Grid Reliability*: Grid reliability is under serious challenge with the integration of renewable resources, demand response, and load management mechanisms introduced with smart grid migration. Added to this are the storage devices and the plugged hybrid electric vehicles demanding quick response from the grid, with the aid of a communication infrastructure for maintaining grid reliability.

3.16   Summary

This chapter provides insights into the communication components of the SCADA systems, which act as the vital link between the RTU/IED and the master station. The chapter starts with the SCADA and smart grid communication requirements and discusses the network topologies and techniques used. Data communication is dealt with in detail, discussing the transmission of digital signals, modes of data communication, and the error detection and media access control (MAC) techniques used in SCADA communication. The SCADA communication protocol general architecture is discussed with OSI, EPA, and TCP/IP models. The evolution of SCADA communication protocols is presented, and all the important SCADA and smart grid protocols like IEC 60870-5, DNP 3, Modbus, ICCP, IEC 61850, IEEE C37.118, and ZigBee architecture are discussed. A detailed discussion of the communication media used, both guided and unguided, follows this. The chapter concludes with discussion of the security of SCADA communication and the challenges ahead.

References

1. Ye Yan, Yi Qian, Hamid Sharif, and David Tipper, A survey on smart grid communication infrastructures: Motivations, requirements and challenges, *IEEE Communications Surveys and Tutorials*, vol. 15, no. 1, pp. 5–20, first quarter 2013.
2. John D. McDonald, *Electric Power Substations Engineering*, 3rd ed., CRC Press, Boca Raton, FL, 2012.
3. IEEE Tutorial Course on Advancements in Microprocessor-Based Protection and Communication, IEEE Power Engineering Society Piscataway, NJ, 97TP120,.
4. Donald J. Marihart, Communications technology guidelines for EMS/SCADA systems, *IEEE Transactions on Power Delivery*, vol. 16, no. 2, pp. 181–188, April 2001.
5. Behrouz A. Forouzan, *Data Communications and Networking*, 4th ed., McGraw-Hill, New York, 2006.
6. Gordon Clarke and Deon Reynders, *Modern SCADA Protocol, DNP3, IEC 60870-5 and Related Systems*, Elsevier, Amsterdam, The Netherlands, 2008.
7. Cobus Strauss, *Electric Network Automation and Communication Systems*, Newness, Elsevier, Amsterdam, The Netherlands, 2003.
8. Stuart A. Boyer, *SCADA Supervisory Control and Data Acquisition*, 4th ed., International Society of Automation (ISA), Research Triangle Park, NC, 2010.
9. IEEE Standard for Synchrophasor Measurement for Power System, IEEE std C37.118.1™-2011 [Online]. Available: ieeexplore.ieee.org.
10. IEEE Standard for Synchrophasor Data Transfer for Power System, IEEE std C37.118.2™-2011 [Online]. Available: ieeexplore.ieee.org.

11. D. -M. Han and J. -H. Lim, Smart home energy management system using IEEE 802.15.4 and ZigBee, IEEE Transactions on Consumer Electronics, vol. 56, no. 3, pp. 1403–1410.
12. Khusvinder Gill, Shuang-Hua Yang, Fang Yao, and Xin Lu, A ZigBee-based home automation system, *IEEE Transactions on Consumer Electronics*, vol. 55, no. 2, pp. 422–430, May 2009.
13. Carles Gomez and Josep Paradells, Wireless home automation networks: A survey of architectures and technologies, *IEEE Communications Magazine, Consumer Communication and Networking*, June 2010.
14. John Walko, Home control, *IET Computing and Control Engineering*, October/November 2006.
15. Julie Hull, H. Khurana, T. Markham, and K. Staggs, Staying in control, *IEEE Power and Energy Magazine*, pp. 41–48, January/February 2012.
16. Mini S. Thomas and Iqbal Ali, Reliable, fast and deterministic substation communication network architecture and its performance simulation, *IEEE Transactions on Power Delivery*, vol. 24, no. 4, pp. 2364–2370, October 2010.
17. Mini S. Thomas, Remote control, *IEEE Power and Energy Magazine*, vol. 8, no. 4, pp. 53–60, July/August 2010.
18. Ikbal Ali, Mini. S. Thomas, and Sunil Gupta, Substation communication architecture to realize the future smart grid, *International Journal of Energy Technology and Policies*, vol. 1, no. 4, 2011.
19. Iqbal Ali and Mini S. Thomas, Substation communication networks architecture, *IEEE POWERCON-2008*, New Delhi, India, October 2008. DOI 10.1109/IC PST 200, 8.474521.8.
20. Iqbal Ali, Mini S. Thomas, Ethernet enabled fast and reliable monitoring protection and control of electric power substation, *IEEE International Conference, PEDES*, New Delhi, India, 2006. DOI 10.1109/PEDES. 2006. 344314.

chapter four

Substation automation (SA)

4.1 Substation automation: Why? Why now?

Substation automation is gaining momentum throughout the world and utilities are rapidly automating the substations due to compelling reasons and to reap all the benefits offered by the new devices and systems.

Substation automation involves the deployment of substation and feeder operating functions and applications ranging from supervisory control and data acquisition (SCADA) and alarm processing, to integrated volt-var control in order to optimize the management of capital assets and enhance operation and maintenance (O&M) efficiencies with minimal human intervention.

The following sections elaborate the need for automating the substations and explain why the time is ripe for the much needed overhaul of the systems.

4.1.1 Deregulation and competition

In a deregulated market, utilities are competing with each other and selling electricity directly to the customers. Deregulation has created competition among the supplying utilities, which has led to lower prices and a chance for customers to find the best deal.

The competition has also improved the power quality, service reliability, and cost of service, as expected. New energy-related services and business areas allow utilities to invest more in automating the substations. The availability of various kinds of information from the system which improves decision making has also made the utilities proactive toward substation automation.

4.1.2 Development of intelligent electronic devices (IEDs)

The rapid development and deployment of IEDs has boosted the substation automation (SA) business as it has opened up new opportunities. Protective relays, meters, and equipment condition monitoring IEDs have been installed and are integral parts of many substations. The technological developments have made the IEDs and the SA systems more powerful, at the same time less expensive, and utilities can justify the investment.

4.1.3 Enterprise-wide interest in information from IEDs

The IEDs, as discussed in Chapter 2 have unmatched capabilities as far as the data capture from the field is concerned. The operational data, consisting of the current, voltages, watts, VARs, fault location, switch gear status, and so forth, are available to the personnel in the entire utility when required. In addition, the IEDs provide nonoperational data, which include fault event (waveform) and power quality data that are vital for post-event analysis and decision making. The beauty lies in the fact that the data can also be accessed by personnel working outside the control room for better planning and decisions for the future.

4.1.4 Implementation and acceptance of standards

As discussed in Chapter 3, the earlier marketplace was full of proprietary communication protocols, and it was difficult for devices from different vendors to communicate. However, the confusion over the communication standard is diminishing, and many international standards like IEC 61850, IEEE 1815(DNP3), and IEC 60870 have become a reality. Standards-based implementations of projects are underway throughout the world. Hence, it has become easier and simpler for utilities to implement automation of substations and related distribution systems.

4.1.5 Construction cost savings and reduction in physical complexity

The functionality bundling of devices has resulted in fewer components and has reduced the number of devices to be purchased, commissioned, and implemented. One IED may replace many electromechanical devices, and there will be construction cost savings. Since the number of devices reduces drastically, there will be less inter-device wiring, and some traditional devices will be eliminated altogether. Relay panel and control house size will be reduced and design and construction costs are also reduced considerably.

All of the above factors have contributed to the acceptance of substation automation as a necessity, and more personnel are being trained and new projects implemented world over. It is imperative to look at the conventional and modern substations to better understand the concepts.

4.2 Conventional substations: Islands of automation

Conventional substation design encompasses the high-voltage switchgear with copper cables interfacing the primary and secondary equipment.

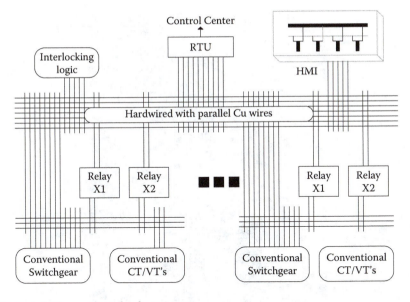

Figure 4.1 Conventional substation.

The substation control room has a number of relays and other devices for protection, SCADA, metering, and so on. They required more control room space as well as more devices. The analog voltage and current signals and the binary switch positions reach the control room via copper cables. Figure 4.1 presents a conventional substation and the wiring requirements.

Cables of different lengths and sizes are used in a substation for carrying the signals, depending on the location of the switchgear and complexity of the control and protection system. It may be noted that a typical substation will have multiple instrument transformers for protection, control, and measurement purposes. Cables are cut to specific lengths and bundled together, as given in Figure 4.2, and any future modification is very labor intensive, especially in older substations where the cabling is starting to deteriorate. The large distances between the devices and control room expose the cabling to electromagnetic interferences and damages. The resistance of the cables also becomes important when selecting the instrument transformers and protection equipment. CT saturation plays a vital role in protection relays operating under fault conditions.

In a conventional substation, there were distinct islands of automation as shown in Figure 4.3. SCADA systems had a dedicated set of devices, including instrument transformers doing the monitoring and control of the system, mostly utilizing hardwire, as discussed earlier. Metering used a separate set of devices to ensure accuracy. The instrument transformers with iron cores were designed for different functionalities. The protection

Figure 4.2 Panel with bundles of wiring. (Courtesy General Electric.)

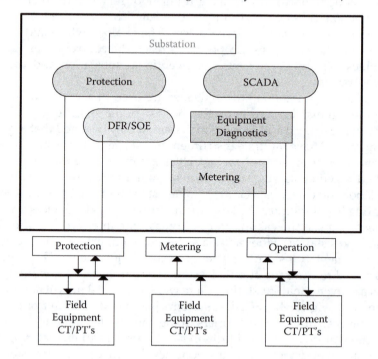

Figure 4.3 (See color insert.) Islands of automation in a conventional substation.

functions received priority, and had CTs and PTs directly connected to the relays with dedicated hardwired channels from relays to the trip coils of the circuit breakers. Digital fault recorders were installed to record the fault waveforms for further analysis and evaluation.

When the substations became automated, the islands started merging, and a need was felt to eliminate duplication and hardwiring while migrating to digitized substations. At the same time, there was a need for the individual groups who maintained only their equipment in the substation to work closely with other groups. There was also a need to define a new skill set of a "super IED specialist" that combined the skills of remote terminal unit (RTU) support, protection, and communications into one position.

An array of new devices and technologies has revolutionized the substation design and the following section elaborates a few of them. It may be noted that the monitoring has become "intelligent" with the new instrument transformers for measurement and circuit breakers for control.

4.3   New smart devices for substation automation

It is evident that substation automation is implemented to reduce human intervention and to improve the operating efficiency of the system. Optimization of assets and reduction of operating costs in the long run are added advantages of the substation automation. The substation and feeder operating functions will be discussed in detail in this chapter and in subsequent chapters.

It is apparent that a set of new devices has been developed and implemented in the substation, which is making the substations intelligent. The following are some of these smart devices.

4.3.1   IEDs

IEDs are key components of substation integration and automation technology. Substation integration involves integrating protection, control, and data acquisition functions into a minimal number of platforms to reduce capital and operating costs, reduce panel and control room space, and eliminate redundant equipment and databases.

IEDs facilitate the exchange of both operational and nonoperational data. Operational data, also called supervisory control and data acquisition (SCADA) data, are instantaneous values of power system analog and status points such as volts, amps, MW, MVAR, circuit breaker status, and switch position. These data are time critical and are used to monitor and control the power system (e.g., opening circuit breakers, changing tap settings, equipment failure indication, etc.). Nonoperational data consist of files and waveforms such as event summaries, oscillographic event reports, or sequential events records, in addition to SCADA-like points (e.g., status and analog points) that have a logical state or a numerical value. These data are not needed by the SCADA dispatchers to monitor and control the power system.

As discussed earlier, IEDs are revolutionizing the substation, and in Chapter 2, the functionalities have been discussed in detail. The IEDs combine protection, metering, control, and automation functions into a single device, thus simplifying the new substation design.

4.3.2   New instrument transformers with digital interface

New instrument transformers are available in the market. They can be directly linked to a merging unit, which in turn sends digital data over a network to the protection and metering devices and can eliminate the hardwiring to a large extent.

New coreless instrument transformers are revolutionizing the substation automation scenario, with protection and metering capabilities. The root cause of many limitations of conventional instrument transformers is their reliance on an iron core. The core is a source of inaccuracy, due to the need to magnetize it, but at the same time should not cause saturation. In the case of conventional CTs, achieving the low-level accuracy and dynamic range to satisfy both measurement and protection duties is a challenge. Conventional VTs may experience ferroresonance phenomena resulting in thermal overstressing. The new instrument transformers use capacitive, optical, and Rogowski techniques to capture the voltage and current from the field, thus making the systems robust, smaller, and reliable.

Optical sensors use the Faraday effect, wherein a fiber-optic loop sensor carrying a polarized light beam encircles the power conductor. Due to the magnetic field created by the primary current flow, the light will experience an angular deflection. Based on the real-time optical measurement, the primary current is accurately detected.

In Rogowski sensors, four quadrants of a multilayer printed circuit board are clamped together to form a toroid around the primary conductor. The sensor output becomes a low-level voltage measurement that can be accurately correlated to the primary current.

Capacitive dividers are replacing conventional core voltage transformers. Printed circuit boards are used on the enclosures of gas-insulated substations (GIS), and slim-line film stacks are used for air-insulated substations. The outputs of the instrument transformers are connected to the merging unit, which sends the digitized signals to the substation automation higher hierarchy.

Nonconventional instrument transformers with digital interfaces based on IEC 61850-9-2 (process bus), as discussed in Section 4.6.3, result in further improvements and can help eliminate some of the issues related to the conflicting requirements of protecting and metering IEDs.

4.3.3   Intelligent breaker

Intelligent breaker has a digital interface that can access digital data from a local area network (LAN) and take action accordingly. It can also transmit back information, especially status changes and other data, through the LAN. An intelligent breaker has a controller inside which can be programmed to make appropriate decisions as per the system conditions.

4.3.4   Merging units (MUs)

The interface of the instrument transformers (both conventional and non-conventional) with different types of substation protection, control, monitoring, and recording equipment is through a device called a *merging unit*. This is defined in IEC 61850-9-1 as follows: "Merging unit: interface unit that accepts multiple analogue CT/VT and binary inputs and produces multiple time synchronized serial unidirectional multi-drop digital point to point outputs to provide data communication via the logical interfaces 4 and 5."

The existing merging units have the following functionalities:

- Signal processing of all sensors—conventional or nonconventional
- Synchronization of all measurements—three currents and three voltages
- Analogue interface—high- and low-level signals
- Digital interface—IEC 60044-8 or IEC 61850-9-2

The merging units should be able to interface with both conventional and nonconventional sensors in order to allow the implementation of SA in existing or new substations.

The merging unit is similar to an analog input module of a conventional protection device or another multifunction IED. The difference is that in this case the substation LAN performs as the digital data bus between the input module and the protection or other functions in the devices. The units are located in different devices, representing the typical IEC 61850 distributed functionality.

4.4   The new integrated digital substation [12,3,8,10,13]

Integration is the key to a new substation, where the islands of automation as depicted in Figure 4.3 vanish, and a common platform is created for protection, monitoring, metering, and many more functions in a substation, as shown in Figure 4.5.

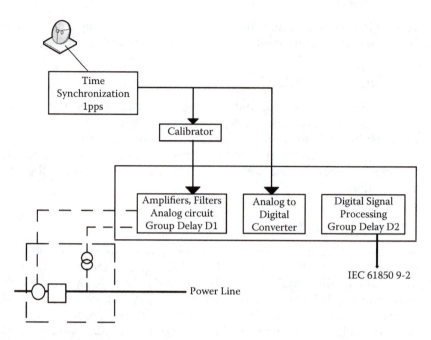

Figure 4.4 Merging unit (MU).

The data acquired from the field are carried to the substation LAN by the merging units (Figure 4.4) and IEDs from where the appropriate applications can acquire data through a common communication channel. The applications are SCADA/SA, metering, equipment diagnosis, waveform analysis, and so forth, and the protection functions are integrated into the SA system, as will be discussed later in Section 4.9.

4.4.1 Levels of automation in a substation

Substation integration and automation can be broken down into three levels of activity, as shown in Figure 4.6. First there is the power system equipment such as transformers, circuit breakers, and intelligent instrument transformers and sensors. The first level is IED implementation where different IEDs are installed in the substation. The second level is IED integration, utilizing the two-way communications of the IED. IEDs from different vendors and with different functionalities have to be integrated to form a cohesive protection, monitoring, and control system, which also performs a number of other functions like waveform recording and metering. Once the IEDs are integrated, a number of substation automation applications, the third level, can be run to effectively monitor and control the substation and associated feeder and customer automation functions in the power system. Last, there is the utility enterprise

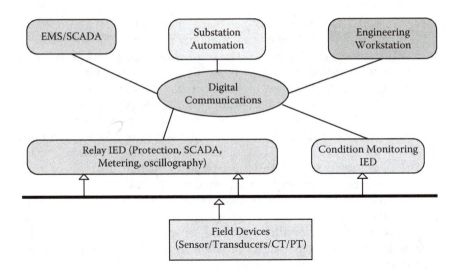

Figure 4.5 (See color insert.) Integration of functions in a substation.

	Utility Enterprise
Level III	Substation Automation functions
Level II	IED Integration
Level I	IED Implementation
	Power system equipment (transformers, breakers, etc.)

Figure 4.6 Levels of substation automation.

where the integration of different control centers can be performed and utility-wide data sharing and applications can be run.

4.4.2 Architecture functional data paths

There are three primary functional data paths from the substation to the utility enterprise: operational data to the SCADA master, nonoperational data to the enterprise data warehouse, and remote access to the IEDs. The most common data path conveys the operational data to the utility's SCADA system at the scan rate of the master (every 2 to 4 s). Operational data are instantaneous values of volts, amps, MW, MVAr, and so forth, which are typically conveyed to the SCADA system using an industry standard communication protocol such as DNP3. This information is critical for the utility's dispatchers to monitor and control the power system. They ensure SCADA dispatchers can effectively monitor and control the system. This operational data path from the substation to the SCADA system is continuous. The most challenging data path is conveying the nonoperational data to the utility's enterprise data warehouse. Nonoperational data are used for analysis and historical archiving and are not in the same "single point" format as operational data; for example, the fault event logs, metering records, and oscillography data are more difficult to extract from the IED relays because the vendor's proprietary ASCII commands are required for nonoperational data file extraction.

The operational data path to the SCADA system utilizes the communication protocol presently supported by the SCADA master. The nonoperational data path to the data warehouse conveys the IED nonoperational data from the substation automation (SA) system to the data warehouse. The data are pulled by a data warehouse application from the SA system or pushed from the SA system to the data warehouse based on an event trigger or time. The remote access path to the substation utilizes a dial-in telephone or network connection to transmit data from a remote IED to be assessed by a substation field worker from the office. See Figure 4.7 for an illustration of the functional data paths.

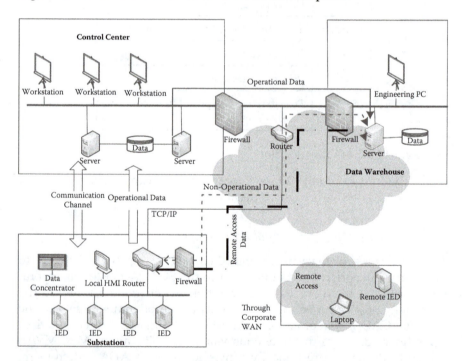

Figure 4.7 (See color insert.) Functional data paths (operational, nonoperational, and remote access).

4.4.3 Data warehouse

In order to make full use of the data available in a utility substation, the concept of data warehousing was introduced in the power sector. A data warehouse is a subject-oriented, integrated, time-variant, and nonvolatile collection of data in support of a manager's decision-making process. Because data are of vital importance in all kinds of applications, it is appropriate to have a data center available for exchange, management, and utilization of data. Data warehouses in general are useful as they help in finding the patterns, trends, facts, relations, models, and sequences hidden in the raw data of the operational environment, to enable better decision making for the optimum operation of the system.

A power system is mainly a real-time industrial system that consists of a primary system responsible for the generation, transmission, and distribution of electricity, and a secondary system in charge of the supervision, control, and operation of the primary system. The secondary system is where computers are most widely used, and this may be a central supervisory control and data acquisition (SCADA) system thereby requiring a data warehouse.

The corporate- or enterprise-level data warehouse or data mart is a server or group of servers that retrieve data from the local data marts, which typically are linked to a system such as SCADA, substation automation, power plant distributed control, maintenance management, and customer information systems. The corporate data warehouse accesses and stores these files centrally and integrates the datasets into unique information that is delivered to, or accessed when needed by, specific user groups in engineering, operations, and maintenance. See Figure 4.9.

Most of the automated systems have local archives, known as historians, already built in. Typically, the local historians are not designed for data warehouse integration, to push data to a central data warehouse or to pull data when required by the central data mart.

The overall objective of the data warehouse technology is to harness and integrate this valuable data, process the data into useful information, and serve the data to applications and personnel for analysis at all levels, as depicted in Figure 4.8. When accurate, timely information regarding the performance of systems and equipment is available to personnel throughout the enterprise, management can start making better decisions, which benefits the organization. In addition, the utility is able to maintain assets more efficiently and effectively by planning equipment upgrades and realizing more life spans for aging components. The data mart eliminates duplication of data residing in multiple databases in the personal computers of utilities.

Data warehousing enables users to access substation data while maintaining a firewall to substation control and operation functions. Both

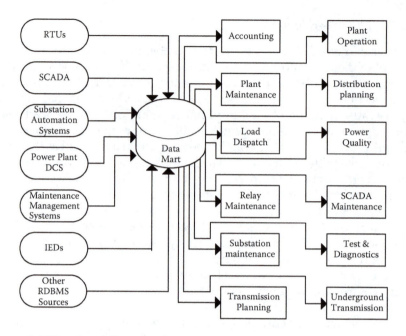

Figure 4.8 The virtual data warehouse vision.

operational and nonoperational data are needed in the data warehouse. To size the data warehouse, the utility must determine who the users of the substation automation system data are, the nature of their application, the type of data needed, how often the data are needed, and the frequency of update required for each user.

The creation of the data warehouse is as follows. The nonoperational data from IEDs or computer systems (such as an energy management system [EMS] or an outage management system [OMS]), as shown in Figure 4.7, are sent upstream to a data concentrator at the substation. These data then travel across the corporate firewall to the corporate side to be stored in a manner that allows queries and data mining by business units on the corporate network. Operational data have been routed from the substation to the control center. A subset of this operational data also is sent to the corporate side for access by business units. Enterprise personnel need operational data to augment their findings, and they can access the data on the corporate side. The mode of storage is not important. What is crucial is that the enterprise must create a data mart that allows enterprise business groups to query or mine all nonoperational data for their myriad purposes. To achieve full value by its users, a data mart must provide the results of queries and mining in useful form to the end user. So the design of an informational architecture in this application must begin by addressing who needs the data, which data must be accessed,

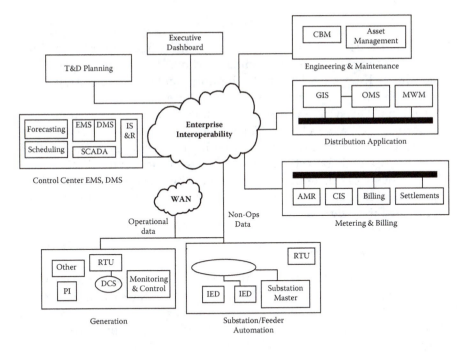

Figure 4.9 The enterprise interoperability concept.

in what form, and in what specific temporal intervals. (These steps echo a utility's approach to distribution automation in general. Where are IEDs placed and what metrics are needed?) Polling (and prodding) all enterprise stakeholders on their data needs leads to an enterprise-wide data requirements matrix, a map that connects data needs with data sources. This process then leads to an inventory of IEDs in the field and their data attributes. Matching the enterprise data map on the need side with the inventory of the IEDs that provide pertinent data on the system completes the picture, at least conceptually. In addition to FEPs, other data sources in the enterprise are shown in Figure 4.9.

The functionalities of a data warehouse are discussed in detail later in this chapter.

4.5   Substation automation: Technical issues

Automating a substation reveals many technical issues where critical decisions have to be taken at the utility level. The investment requirements play a major role in the final automation plan of the substation; however, there are certain aspects that cannot be compromised while making decisions, as discussed below.

4.5.1   System responsibilities [1,8,23]

The system must interface with all of the IEDs in the substation. This includes polling the IEDs for readings and event notifications. The data from all the IEDs must be sent to the utility enterprise to populate the data warehouse or be sent to an appropriate location for storage of the substation data. The system processes data and control requests from users and from the data warehouse. The system must isolate supplier proprietary functionality by providing a generic interface to the IEDs. In other words, there should be a standard interface regardless of the IED supplier. The system should be updated with a report-by-exception scheme, where status-point changes and analog-point changes are reported only when they exceed their significant deadbands.

4.5.2   System architecture

The types of data and control that the system will be expected to facilitate are dependent on the choice of IEDs and devices in the system. This must be addressed on a substation-by-substation basis. The primary requirement is that the analog readings be obtained in a way that provides an accurate representation of their values:

- *Level 1 Field Devices:* Each electronic device (relay, meter, PLC, etc.) has internal memory to store some or all of the following data: analog values, status changes, sequence of events, and power quality. These data are typically stored in a FIFO (first in, first out) queue and vary in the number of events maintained.
- *Level 2 Substation Data Concentrator:* The substation data concentrator should poll each device (both electronic and other) for analog values and status changes at data collection rates consistent with the utility's SCADA system (e.g., status points every 2 s, tieline and generator analogs every 2 s, and remaining analog values every 2 to 10 s). The substation data concentrator should maintain a local database.
- *Level 3 SCADA System/Data Warehouse:* All data required for operational purposes should be communicated to the SCADA system via a communication link from the data concentrator. All data required for nonoperational purposes should be communicated to the data warehouse via a communication link from the data concentrator. A data warehouse is necessary to support a mainframe or client-server architecture of data exchange between the system and corporate users over the corporate WAN (wide area network). This setup pro-

vides users with up-to-date information and eliminates the need to wait for access using a single line of communications to the system, such as telephone dial-up through a modem.

4.5.3 Substation host processor

The substation host processor must be based on industry standards and strong networking ability, such as Ethernet, TCP/IP, Windows operating system, Linux, and so forth. It must also support an open architecture, with no proprietary interfaces or products. An industry-accepted relational database (RDB) with structured query language (SQL) capability and enterprise-wide computing must be supported. The RDB supplier must provide replication capabilities to support a redundant or backup database.

4.5.4 Substation LAN

The substation LAN must meet industry standards to allow interoperability and the use of plug-and-play devices. Open-architecture principles should be followed, including the use of industry standard protocols (e.g., IEEE 802.x [Ethernet]). The LAN technology employed must be applicable to the substation environment and facilitate interfacing to process-level equipment (IEDs, PLCs) while providing immunity and isolation to substation noise.

4.5.5 User interface

The user interface in the substation must be an intuitive design to ensure effective use of the system with minimal confusion. An efficient display hierarchy will allow all essential activities to be performed from a few displays. It is critical to minimize or, better yet, eliminate the need for typing. There should be a common look and feel established for all displays. A library of standard symbols should be used to represent substation power apparatus on graphical displays. In fact, this library should be established and used in all substations and coordinated with other systems in the utility, such as the distribution SCADA system, the energy management system, the geographic information system (GIS), and the trouble call management system.

4.5.6 Communications interfaces

There are interfaces to substation IEDs to acquire data, determine the operating status of each IED, support all communication protocols used by the IEDs, and support standard protocols being developed. There may be an interface to the energy management system (EMS) that allows system operators to monitor and control each substation and the EMS to receive data from the substation integration and automation system at different periodicities. There may be an interface to the distribution management system with the same capabilities as the EMS interface.

4.5.7   Protocol considerations

As discussed earlier, a communication protocol allows communication between two devices. The devices must have the same protocol (and version) implemented. Any protocol differences will result in communication errors. The substation integration and automation architecture must allow devices from different suppliers to communicate (interoperate) using an industry-standard protocol. The utility has the flexibility to choose the best devices for each application, provided the suppliers have designed their devices to achieve full functionality with the protocol. There are two capabilities a utility considers for an IED. The primary capability of an IED is its stand-alone capabilities, such as protecting the power system for a relay IED. The secondary capability of an IED is its integration capabilities, such as its physical interface (e.g., RS-232, RS-485, Ethernet) and its communication protocol (e.g., DNP3, Modbus, IEC 61850 MMS). Utilities typically specify the IEDs they want to use in the substation rather than give a supplier a turnkey contract to provide the supplier's IEDs only in the substation. However, utilities typically choose the IEDs based on the IED's stand-alone capabilities only, without considering the IED's integration capabilities.

4.6   The new digital substation

The new digital substation will have smart devices and smart wiring with smart controllers making it a designer's dream, as it will have integrated protection, metering, data retrieval, monitoring, and control capabilities.

The substation is organised in three architectural levels.

4.6.1   Process level

Conventionally the data from the equipment in a substation reach the devices for processing via hardwire. In the modern substation, the equipment (primary systems) in the substation is embedded with smart sensors, which can directly communicate over a LAN, which is referred to as a process bus as shown in Figure 4.10. The current, voltage, and other data (pressure and temperature from GIS, etc.) from the field are transmitted

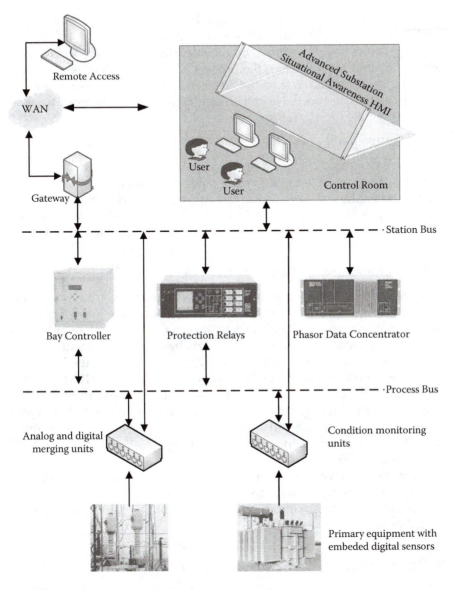

Figure 4.10 (See color insert.) The new digital substation.

over the Ethernet process bus. The devices, such as protection relays, recorders, PMUs, bay controllers, and PDCs, can readily subscribe to this information as clients to this data flow over the process bus. The information from the field devices (primary equipment) travels to the SCADA control center via the process bus, and the control signals from the control station reach the field devices through the process bus. In a fully digital architecture, control commands are also routed to the primary devices via the process bus. The process bus enables time-critical services for both protection and control. Thus, the process bus aids in effectively communicating the information from the "eyes and ears" of the power system to the bay level more effectively and efficiently than the conventional means.

4.6.2 Protection and control level

There are devices in a substation which link the process bus with the station bus. These include the secondary equipment like bay controllers, protection relays, Ethernet network and switches, time synchronization units, measuring devices, and recording devices. The bay level or the protection and control level include these devices and the panels that host them.

4.6.3 Station bus and station level

The station bus is the LAN that supports peer-to-peer communication and multiple devices and clients to exchange data. The communication of data to higher hierarchy also originates from this layer. The IEDs perform the time-critical protection functions by directly interacting on the process bus. The point-on-wave switching is also done by the IEDs. The station will host the substation human-machine interface (HMI) required to visualize the events in the substation so that the personnel will get the real-time data of the operations. Coordination of multiple IEDs and also the condition monitoring of the equipment in the substation (transformers, circuit breakers, and bushings, etc.) are managed by the station-level HMI via the station bus.

Figure 4.10 gives the picture of the modern digital substation with the equipment in the field (process level based on IEC 61850-9-2), the bay-level devices and communication, and the station-level LAN and the associated devices and databases.

4.7 Substation automation architectures

The substation automation architecture varies depending on the components finalized for deployment in the substation and for automation. The automation of conventional substation and the new substation will differ greatly due to the availability of devices and communication channels. The migration from legacy systems to modern automation systems provides crucial insights into the evolution of the automation system for substations. The following sections elaborate this migration.

4.7.1   Legacy substation automation system

Electric substations are the most critical infrastructure of the electric power systems as they monitor and control the widely spread transmission network. As discussed earlier, automation of the substations is in progress across the world at a rapid pace to equip the utilities to utilize the available resources in the most efficient manner. The development of IEDs, the implementation and acceptance of standards, and migration to interoperable open systems have led to the widespread automation of substations.

Automating a new substation is relatively easy. Work can start at the onset; new IEDs can be implemented and integrated for performing different functions at the design stage. However, automating existing substations poses a serious challenge. Several alternatives are available for implementation depending upon the availability of equipment and software in the substation and the financial constraints. In order to integrate the IEDs with an existing system, it is necessary that the engineers are well versed with the old and new technologies.

The traditional substation automation system is shown in Figure 4.11 with the substation receiving data from the hardwired inputs from the transducers and other devices in the field for operational data. These are transmitted to the SCADA system at higher hierarchical levels. Protection relays also report to the substation. The substation field and the bay level will have a large amount of hard copper wiring, and any expansion of the system will require extensive trenching and wiring and correcting the related difficulties.

The migration of a legacy substation to a digital substation may be done in stages, as discussed in Section 4.8 on new versus existing substations. When IEDs with additional functionalities are added to the substation to replace the electromechanical relays, the substation will resemble that depicted in Figure 4.12. Still hardwired inputs and outputs remain in the substation for other functionalities.

The new digital substation architecture is discussed in detail in the following section. The migration paths from a nonautomated system to a non–IEC 61850 system and finally to a fully digital substation are discussed.

4.7.2   Digital substation automation design

The new digital substation architecture can be designed in multiple ways. The IEC 61850 may be implemented gradually by starting with adaptation of existing IEDs to support the new communications standard over the station bus and at the same time introducing some first process bus-based solutions. Two types of station architectures are discussed in the following sections.

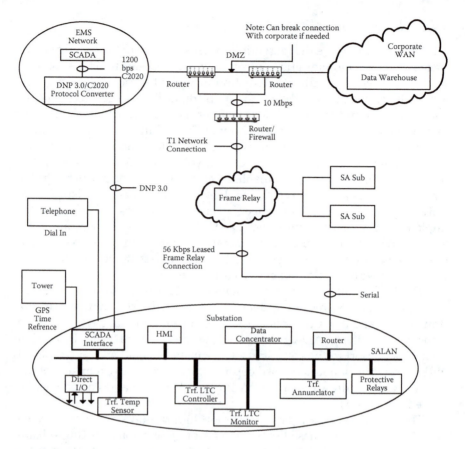

Figure 4.11 Conventional substation.

4.7.2.1   *Station bus architecture [2]*

The functional hierarchy of a station bus-based architecture is shown in Figure 4.13. It represents a partial implementation of IEC 61850 and brings only some of the benefits that the new standard offers. The current and voltage inputs of the devices (protection, control, monitoring, or recording [PCMR]) at the bottom of the functional hierarchy are conventional and wired to the secondary side of the substation instrument transformers using copper cables.

This architecture still offers some significant advantages compared to conventional hardwired systems. It allows for the design and implementation of different protection schemes that in a conventional system require a significant number of cross-wired binary inputs and outputs.

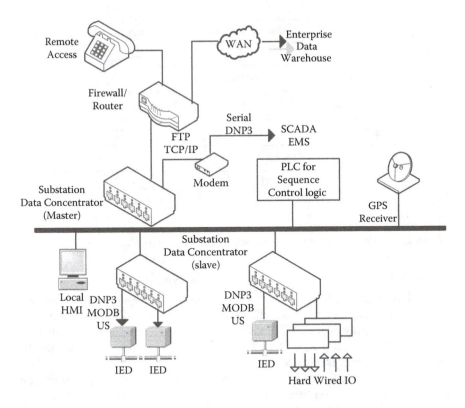

Figure 4.12 Conventional substation with hierarchical architecture.

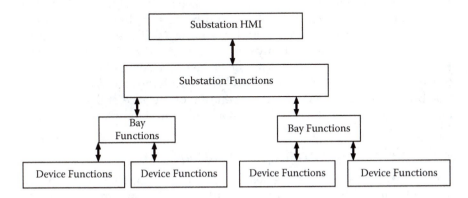

Figure 4.13 Station bus–based architecture.

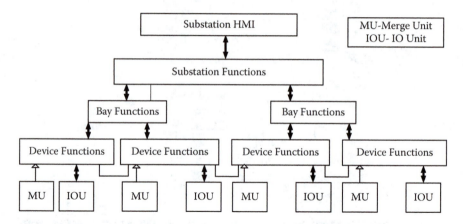

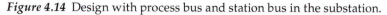

Figure 4.14 Design with process bus and station bus in the substation.

4.7.2.2 Station bus and process bus architecture

Full advantage of all the features available in the new communications standard can be taken if both the station and process bus are used. The IEC 61850 communications-based distributed applications involve several different devices connected to a substation LAN as shown in the simplified block diagram in Figure 4.14.

A merging unit (MU) will process the sensor inputs, generate the sampled values for the three-phase currents and voltages, format a communications message, and multicast it on the substation LAN.

Another device, the IO unit (IOU), will process the status inputs, generate status data, format a communications message, and multicast it on the substation LAN. All multifunctional IEDs will then receive sampled values messages and binary status messages. The units that have subscribed to this data then process the data (including re-sampling in most of the cases), make a decision, and operate by sending a GSE message to the IOU to trip the breaker or perform any other required action. Figure 4.13 shows the simplified communications architecture of the complete implementation of IEC 61850. The number of switches for both the process and substation buses can be more than one, depending on the size of the substation and the requirements for reliability, availability, and maintainability.

4.8 New versus existing substations

The design of new substations has the advantage of starting with a blank sheet of paper. The new substation will typically have many IEDs for different functions, and the majority of operational data for the SCADA system will come from these IEDs. The IEDs will be integrated with digital two-way communications. Typically, there are no conventional remote terminal units (RTUs) in new substations. The RTU functionality is addressed using IEDs, PLCs, and an integration network using digital communications. Previously, 100% of the points in a substation were hard-wired, and conventional RTUs were used to process the hardwired input and output (I/O) points. In a new substation, fewer than 5% of the points are hardwired, so conventional RTUs have been replaced by data concentrators that are designed to support both operational and nonoperational data from IEDs.

In existing substations, there are several alternative approaches, depending on whether a substation has a conventional RTU installed. The utility has three choices for their existing conventional substation RTUs: integrate RTU with IEDs (assuming the RTU supports this capability); integrate RTU as another IED; or retire RTU and use IEDs and data concentrator, as with a new substation.

As utilities upgrade the legacy equipment on their distribution systems, an opportunity arises to capture more value from the resulting hybrid configuration of new and existing technology. Utilities are refreshing the basic equipment of the substation, and they also are upgrading its automation gear. The focus is on the addition of intelligent electronic devices (IEDs) to the existing array of remote terminal units (RTUs) and data concentrators, both in the substation and downstream on distribution feeders, and the resulting opportunity to capture more value from the resulting data.

The implementation of IEDs provides not only a rich new source of data that can benefit the entire utility organization, it also improves the business case for IEDs and, not incidentally, aids in eliminating the silos that keep a utility's operations staff and information technology (IT) staff from full cooperation.

4.8.1   Drivers of transition

Microprocessor-based IEDs with two-way communications capabilities are augmenting analog RTUs because the former provide much greater functionality than the latter. Specifically, RTUs provide operators only with operational data. IEDs provide operational data and nonoperational data with high potential value to most if not all utility business units. Using only the operational data from an IED, which costs an average of $5,000, is tapping only some 20% of its potential value. Accessing IEDs' nonoperational data can provide business units with insights that will reap benefits across the board, from planning and engineering to maintenance, asset management, and power quality groups—the gamut of utility enterprise entities.

4.8.2 Migration paths and the steps involved

The substation upgrade is a series of steps or levels of activity. First comes the addition of IEDs, which support the three functional data paths—operational, nonoperational, and remote access—taking data upstream to operators in the control center and to the enterprise data mart. Second is the integration process, which takes advantage of the two-way communications capabilities in IEDs for their operational data and recording of nonoperational data to tap the other 80% of their value. Third, the utility must consider which applications to run at the substation level to optimize the operation of the substation and downstream, distribution feeders. Initially, IEDs were integrated with some legacy RTUs, but that arrangement still fell short of delivering the full value of such an upgrade. But the development of a family of products known as data concentrators enabled the collection and transmission of nonoperational data from the IEDs to the enterprise—the key to greater value described here. (Operational data, meanwhile, are routed to the control center.) The result of this transition from an RTU-centric arrangement to a distributed network architecture with IEDs and data concentrators is in contrast to the legacy, serial, point-to-point communications-based architecture. With the distributed network architecture, the legacy RTUs, if they can be maintained, simply assume the role of an IED until the end of their useful lives. Figure 4.15 shows the migration of a RTU-centric hardwired system to the distributed network architecture with data concentrators.

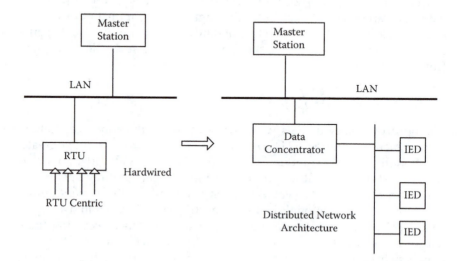

Figure 4.15 Migration from RTU-centric architecture to distributed network architecture.

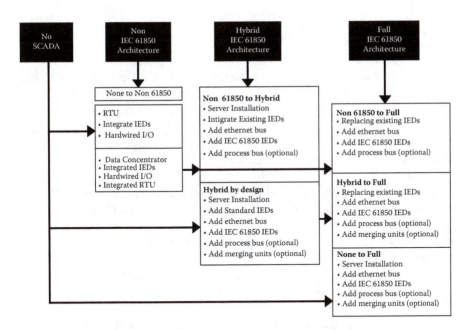

Figure 4.16 Migration paths from no SCADA to legacy, to hybrid, and to full digital substation.

The migration path from a nonautomated substation to a fully digital substation can be achieved in different ways, depending on the availability of resources and the speed at which the migration is desired. Figure 4.16 gives a comprehensive picture of the available migration paths available.

Migrating from a non-SCADA system to a SCADA system begins with installing RTU, IEDs, hardwired I/Os to the RTU and the software interfaces. The migration from no SCADA can also be to a hybrid system as shown in Figure 4.16 where the IEC 61850 IEDs can be added and integrated. Process bus and merging units are optional in this case. These hybrid systems can migrate to the full IEC 61850 functionality easily by replacing the existing IEDs to IEC 61850 compliant ones, adding process bus and merging units. Migration from non-SCADA to legacy to hybrid to full digital substation can be achieved by following any of the mitigation paths, as decided by the utility management.

4.8.3 Value of standards in substation automation [11,12]

The addition of new technology to legacy systems and the resulting hybrid configuration underscores the value of standards. A utility's legacy system is constrained by its current capabilities and the need for an upgrade. A utility must ask: if the legacy vendor remains in business, is the legacy equipment based on open architecture and industry standards, can it be upgraded, are spare parts still available, and is there a logical migration path to the newer technologies with the existing devices?

The historic work of the government-funded Smart Grid Interoperability Panel (SGIP) 1.0 and its transition to a membership-driven SGIP 2.0 laid the foundation for the smart grid standards and interoperability process and global harmonization of related efforts and outcomes. This includes the backward compatibility of IEDs, for instance, and continued use of legacy RTUs as automation is improved.

For instance, IEC 61850, a global standard for substation automation communications, provides the benefit of standard variable nomenclature in place of the unique variable nomenclature of each vendor, endemic to legacy systems. The standardization of terminology regarding technology makes assembling a points list, for instance, much simpler, and eases the transition to automation.

Standards also apply to data extraction, concentration, and storage. These are critical processes that deliver much value of automation to grid operators.

New technologies can be integrated with legacy systems, and that creates a hybrid configuration with a new data network architecture. The integration of old and new is one challenge. But the promise, in the case of substation automation—an important first step in distribution automation in particular and grid modernization in general—is a new distributed network architecture with IEDs and data concentrators that yield high value with the delivery of (operational and) nonoperational data to business units. This improves the return on investment for an expensive proliferation of IEDs. It requires a degree of cooperation between operations and IT, and it brings enterprise-wide value to the utility. Proper integration of new technologies with legacy systems, particularly in the case of substation automation, opens a new, significant value stream in addition to the operational benefits one would expect.

4.9   Substation automation (SA) application functions

As discussed earlier, distribution SCADA will have the basic functions like monitoring and control, report generation, and historical data storage and several functions for special application in the substation automation scheme. The following sections will elaborate on the application functions.

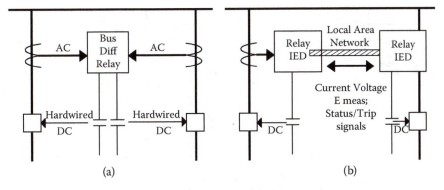

Figure 4.17 (a) Protection via hard wiring and (b) protection via LAN.

4.9.1 Integrated protection functions: Traditional approach and IED-based approach

In the traditional approach, the relays had hardwired inputs from the instrument transformers, and from the relays, hardwires carried the trip signals to the circuit breakers, as shown in Figure 4.17. In the IED-based modern approach, the information from instrument transformers will reach the relay IEDs via LAN, the relays exchange information via LAN, and with the process bus becoming a reality, the circuit breakers will receive a trip signals via a generic object-oriented substation event (GOOSE) message traveling in the process bus, as illustrated in Figure 4.17b.

In case of a breaker failure, in the traditional protection wiring scheme, the hardwiring will carry the trip signal to the backup protection scheme, as shown in Figure 4.18a, while in a modern protection scheme, the backup protection is initiated via LAN as given in Figure 4.18b, which reduces the wiring tremendously and also will utilize alternate pathways.

Protection functions like automatic reclosing and bus differential schemes can be implemented, and breaker failure can be dealt with effectively. The benefit lies in the fact that separate protection relays can be avoided, and thus performance improvement and reliability enhancement are achieved.

4.9.2 Automation functions

The substation automation application functions include intelligent bus failover, automatic load restoration, adaptive relaying, and equipment condition monitoring, which are explained in the following sections.

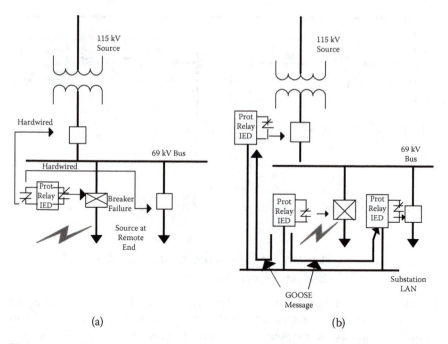

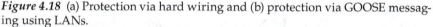

(a)   (b)

Figure 4.18 (a) Protection via hard wiring and (b) protection via GOOSE messaging using LANs.

4.9.2.1 *Intelligent bus failover and automatic load restoration*

This scheme is generally used in a distribution substation where there are two transformers and a normally open bus tie breaker. When a transformer in a substation fails, the simple bus failover scheme transfers the load to the healthy transformer in the substation, which may overload the healthy transformer and lead to another failure; hence, the bus failover schemes have been disabled in some cases. The substation firm capacity is limited by the load (overload) that can be carried by the healthy transformer.

However, in an intelligent bus failover scheme, the substation automation system will ensure that the healthy transformer is not overloaded. This can be done by load shedding one or a few of the outgoing feeders temporarily. These feeders can be supplied from an adjacent substation by closing a tie switch, and the disruption in load can be minimized. The benefit of this scheme is primarily improvement in reliability as the transfer of load is done as quickly as possible. Outage duration can be reduced from 30 min to 1 min. The scheme also allows better equipment utilization. "Intelligent" load transfer will allow higher loading under normal conditions, as the substation firm capacity is limited to the amount of load that can be carried following a single contingency, like a transformer failure.

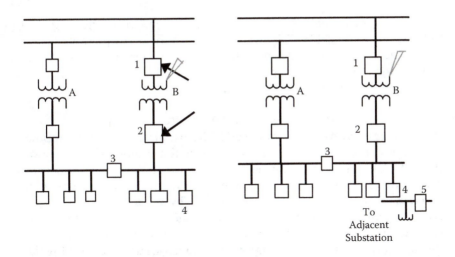

Figure 4.19 Intelligent bus failover demonstration.

Figure 4.19 shows the fault on transformer B. Circuit breakers 1 and 2 will trip and isolate transformer B, and circuit breaker 3 will be closed to transfer the load on to transformer A, which will get overloaded and may have to shed load by opening circuit breaker 4. However, the load can be transferred subsequently by connecting the line to the adjacent substation automatically by closing breaker 5.

4.9.2.2   Supply line sectionalizing

The distribution substations are often tapped off a supply line without high side breaker protection. This creates a problem, and considerable load may be out of service until the field crew arrives. The objectives of the scheme are to identify the faulted section of supply line, isolate the faulted section, and restore supply to the substations fed off the unfaulted section of the supply line. The benefit is that improvement in reliability as the service to substations that are without power can be restored as quickly as possible. The outage duration is reduced from 30 to 1 or 2 min.

4.9.2.3   Adaptive relaying

Adaptive relaying is the process of automatically altering the settings of protective relay IEDs based on the system conditions. This can be of much help to an operator when there is a crisis in the system, with main lines and generators tripping. The special protection function of the substation automation system can play a role by altering the settings of some of the relay IEDs. As an example, in the event of tripping of a critical generator, a line may get heavily loaded due to the diversion of power from other sources. Under normal circumstances, this will lead to the tripping of this line which may escalate to an emergency situation. The master at the control center will detect the trip and can inform the corresponding SA system of the event. The SA system can switch the appropriate relays to the new settings to carry the required power until the crisis is over and then can switch back to the original settings.

4.9.2.4 Equipment condition monitoring (ECM)

Many electric utilities have employed ECM to maintain electric equipment in top operating condition while minimizing the number of interruptions. With ECM, equipment-operating parameters are automatically tracked to detect the emergence of various abnormal operating conditions, using specialized sensors and diagnostic tools. This allows substation operations personnel to take timely action when needed to improve reliability and extend equipment life. This approach is applied most frequently to substation transformers and high-voltage electric supply circuit breakers to minimize the maintenance costs of these devices, by eliminating the time-based routine testing and thus saving significant labor and material cost. The number of catastrophic failures is reduced by reducing the repair cost and avoiding forced outages. Dynamic equipment rating is feasible by which more capacity can be squeezed out of the existing equipment, thus improving the availability. The life of the equipment is thus extended, and even a year or two extension can produce significant savings.

The ECM monitoring devices include

- Dissolved gas in oil monitoring samples
- Moisture detectors
- Load tap changer monitors
- Partial discharge of acoustic monitors
- Bushing monitors
- Circuit break monitors (GIS and OCB)
- Battery monitors
- Expert system analyzers

ECM is an added advantage of substation automation, as discussed later in the chapter.

4.9.3 Enterprise-level application functions

Once the substation is automated, there are many application functions that can be implemented at the enterprise level as discussed in Section 4.4.1. The final level indicates the utility enterprise–related implementations in Figure 4.6.

4.9.3.1 Disturbance analysis

Disturbance analysis is an added advantage when IEDs are implemented, as they have inherent capabilities to record the fault waveform, and also the facility to time stamp the measured operational data. These values can be used to recreate a disturbance sequence as the time stamping to the level of milliseconds or less will help the operator and other personnel in the utility to assess the situation and take corrective action for the next disturbance. For example, in case of an islanding of a section of the power system, the faults, line tripping, overloading and underloading, and tripping of generators may happen in quick succession. After the islanding the utility personnel must assess the situation and determine the correct sequence of events. Time stamping of events, analog, or status changes immensely helps the utility to recreate an event and do a thorough analysis of the situations that led to the islanding. The waveforms captured can be used by the maintenance and protection departments to assess the severity of the damage and take corrective action.

4.9.3.2 Intelligent alarm processing

Intelligent alarm processing is of utmost importance in a control room to help the operator from getting confused by the battery of alarms triggered by an event. This has been explained in detail along with the techniques used in Chapter 2.

4.9.3.3 Power quality monitoring

Power quality is an important factor in today's power system operation scenario for two reasons. The quality of power deteriorates due to the influx of new power electronic devices which flood the system with harmonics and ripples. But quality power is a must for many devices to operate optimally, and a large number of manufacturing facilities require quality power for precision manufacturing. The SA system with the IEDs implemented and integrated can help in the power quality monitoring by reporting the harmonic content in the voltage waveform and the total harmonic distortion and can also send the oscillographic information to the monitoring center for assessment. Necessary corrective action can be taken and the power quality can be maintained by the utility.

4.9.3.4 Real-time equipment monitoring

Real-time equipment monitoring is an offshoot of the equipment condition monitoring discussed earlier. Traditionally, power system equipment is loaded to the rated capacity under normal conditions, whereas if the equipment is condition monitored, the loading can be based on actual conditions, rather than on conservative assumptions. As an example, a transformer detected with a "hot spot" will always be loaded to a much lower value, due to the fear of a catastrophe; however, in a condition monitoring environment, the transformer can be loaded to a higher value by monitoring the true winding hot spot temperature. The loadability can be derived, and it is reported that 5% to 10% additional loading can be achieved by this activity. Hence, the utility can squeeze more capacity out of the existing equipment and delay investment in new equipment.

4.10　Data analysis: Benefits of data warehousing [9,14–22]

The modern substations in the power sector contain large numbers of IEDs as part of the secondary system, each of which captures and stores locally measured analog signals and monitors the operating status of plant items.

Within each substation, the IEDs are networked together via a high-speed LAN capable of transmitting real-time data and control commands. The data exchange between each substation and the control center does not need to consider the minimization of data flow related to the use of a narrow bandwidth communication channel. The open distributed architecture ensures portability of software by implementing a standard application software interface independent of specific hardware platforms. The IEDs and the high-speed communication systems make it possible to convey various analog values (voltage, current, power) and digital signals (circuit breaker status, switch position) to the higher hierarchy as data to the SCADA master station at the scan rate of the system (i.e., 10 s for analog points, 2 s for status points). Such a high scan rate, however, produces so much data that the engineers often suffer from data overload (i.e., the inability to extract knowledge from data). The volume of data burdens traditional information systems in the substation control center and suffocates the existing knowledge acquisition process of expert systems. Thus, improved knowledge extraction techniques are required to assist engineers in analyzing the data received from IEDs and providing quality information to the various user groups. The communication paths to the data warehouse are shown in Figure 4.20.

The major user groups that have been identified in a power utility which will benefit from data warehousing are the operation department, planning department, protection department, engineering department, maintenance department, assets management department, quality department, purchase department, marketing department, safety department, and customer support department.

It is interesting to note that different user groups have different needs regarding the time response and/or extent of information provided by the processed data. The system dispatchers are interested in getting the

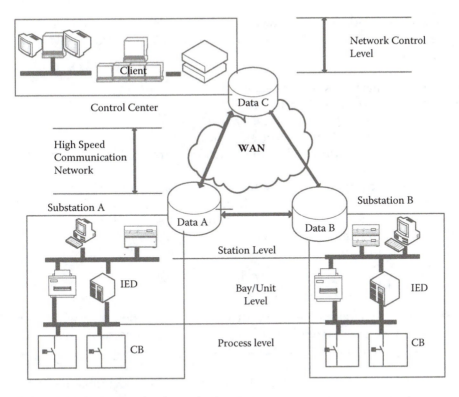

Figure 4.20 Structure of communication system.

condensed fault analysis information as soon as possible after a valid fault occurs. Their main interest is determination of accurate fault location and switching equipment status that enables them to make decisions about the system restoration. The protection engineer, on the other hand, is most interested in getting detailed and specific information regarding the operation of the protection system and related equipment during the event. The time factor in this case is not as strict as for the system dispatcher. Thus, keeping the different needs of different user groups in view, the data communicated at the master station are required to be accurately processed, so that information such as patterns, trends, facts, relations, models, and hidden sequence could be presented to different user groups to achieve better decision making.

4.10.1 Benefits of data analysis to utilities

Data analysis is driven by the latest trends of making the monitoring of power systems more cost effective and focused so that service to the customers is more reliable. The power system operation performance also becomes more competitive. The major benefits of the data analysis are as below:

- Explains why systems behave abnormally
- Restores outages faster
- Prevents problems from escalating
- Operates equipment more efficiently
- Makes informed decisions about infrastructure repair and replacement
- Keeps equipment healthy and extends equipment life
- Improves reliability and availability
- Maximizes the utilization of existing assets
- Improves employee efficiency
- Increases profitability

4.10.2 Problems in data analysis

There are various problems associated with substation data which can generally be categorized into six types and need to be addressed at the data preprocessing or cleaning stage:

- *Redundant Data*: A major achievement of data integration in substations is a high redundancy of data. However, during online data analysis application, this becomes a problem.
- *Irrelevant Data*: The main objective of eliminating irrelevant data is to narrow the knowledge search space.
- *Incomplete or Missing Data*: Missing data can complicate the analysis tasks and hinder the accurate performance of most data analysis systems.
- *Incorrect or Noisy Data*: Nonsystematic errors that occurred during data collection are termed as noise and can pose serious problems analyzing the data.
- *Incompatible Data*: Data compatibility is crucial when data are shared by several workgroups. The IEC61850 standard has been developed to address the incompatibility issue to ensure multivendor IED relays can interoperate over the network.
- *Inconsistent Data*: The inconsistency of data is generally caused by increasing complexities and varieties of products. The synchronization of data using a global positioning system (GPS) clock helps to solve the different frequency of data.

4.10.3   Ways to handle data

The primary task for substation data analysis is to reduce the amount of data, convey a clearer idea of the system condition, and recommend corrective actions to the operators. The following solutions could be considered for optimizing the handling of masses of information:

- *Data Filtering*: The data received at the control center during a typical day do not normally signal any major problems. The data that are not required need to be filtered as they may cause undesirable distraction and confusion.
- *Data Combining*: Data that duplicate information should be combined, so as to help in reduction of redundant information given to the operator.
- *Data Processing*: Data received from the IED relays are usually in an elementary form. The data must be processed before the useful information can be extracted.
- *Data Prioritization*: At the time of an emergency, a very large volume of data is generated. It is imperative to prioritize the available data so as to place the important data in a prominent position. This helps in conveying a clear picture about the system.
- *Data Grouping*: If the data are not processed and delivered properly at the right place and at the right time, the information received by the users may be irrelevant. Thus, it is important to channel different sets of data to different groups on an as-needed basis. This decreases the amount of data required to be handled in a control center.

4.10.4   Knowledge extraction techniques

Knowledge extraction techniques are used in the process of extracting valid, previously unknown, comprehensible, and useful information from large databases and using it. It is an exploratory data analysis, trying to discover useful patterns in data that are not obvious to the user. To provide an implementation framework for the analysis, some of the most common techniques for implementing the automated analysis systems are discussed below. The following techniques used are particularly suitable for a given type of analysis.

Signal Processing: It is obvious that some form of signal processing will take place in almost any analysis implementation that involves analogue waveforms, but some signal processing may be used for the processing of contact status information as well. In the past, the most common signal processing techniques used were those based on orthogonal transforms such as the Fourier transform and its derivatives, fast Fourier transform (FFT) and discrete Fourier transform (DFT).

Expert Systems: The use of expert systems in the implementation of the analysis applications is probably the oldest approach taken. The reason for using this group of techniques is obvious. Analysis is a decision-making process aimed at a number of comparisons and consequent searching steps. Expert system techniques are well suited for that purpose.

Neural Nets: It is well known that neural nets can be a powerful approach to parallel processing of input signals where rather simple and computationally efficient implementation of otherwise complex nonlinear relationships can be achieved. Even though the neural nets have been shown to act as very powerful pattern recognizers, some drawbacks to their use in the analysis are quite serious. As the neural net may have to be extensively trained to become an acceptable classifier, the issue of selection of training sets and methodology needs special attention because different analysis tasks may require different approaches to this issue.

Fuzzy Logic: This set of techniques is often used when dealing with imprecise and/or incomplete data. The theory of defining the fuzzy sets, variables, and logic operations is well known and straightforward. However, applying the theory to the analysis tasks may not be as straightforward because considerable knowledge about the event or device being analyzed is needed to be able to make a selection of the variables and their typical values. For future use of this technique, a better understanding of the benefits and constraints needs to be achieved.

Rough Set: This is a knowledge extraction technique. Rough set theory is used for decision support systems to discover the dependencies that exist within the data, to remove redundancies, and to generate decision rules. One of the major advantages of rough set theory is that it does not require any preliminary or additional information about data, such as grade of membership or the value of possibility in fuzzy set theory.

The main problems that can be approached using rough set theory include data reduction, discovery of data dependencies, estimation of data significance, generation of decision algorithms from data, approximate classification of data, discovery of similarities or differences in data, discovery of patterns in data, and discovery of cause-effect relationships.

Thus knowledge extraction from data provides a whole array of uses for the utility user roles, as given in Figure 4.21. The figure gives an example

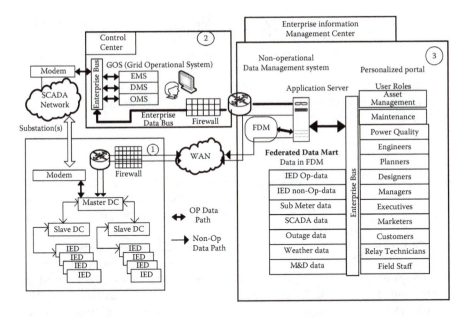

Figure 4.21 The data warehouse for a distribution utility.

of the data mart formation in a distribution substation, where the non-operational data are collated from the IEDs (operational), meter data, outage data, weather data, maintenance data, and SCADA data. Useful information and knowledge from this data is to be distributed to the user groups, asset management, maintenance, protection, power quality, customers, planners, engineers, management, and field staff for making better decisions.

The uses of data warehousing such as disturbance analysis, intelligent alarm processing, and real-time equipment rating were discussed in previous sections.

4.11 SA practical implementation: Substation automation laboratory [4,5,6]

With the development of communication standard IEC 61850, the utilities were forced to implement various options with regard to retrofitting and expansion of existing infrastructure that has not fully completed its operational life, in order to get the maximum benefits of the new technology at the minimum cost.

The SCADA laboratory discussed in Chapter 2 has been assisting Indian engineering students to enhance the practical knowledge and application of SCADA systems. The need to establish a substation automation (SA) laboratory to equip the students with additional knowledge about relay IEDs, communication protocols, and retrofitting of substation equipment was felt in 2008 and its implementation started. Thus, the various modules of the SA laboratory have been designed, keeping in mind the integration of the latest technology available with the existing infrastructure in the substation automation area. These modules help the students to understand the concepts of IEC 61850, interoperability, and the substation migration process.

4.11.1   Hardware design of the SA laboratory

The designing of the substation automation lab has been done with utmost care to incorporate all the components required to demonstrate the capabilities of the IEDs and the standard IEC 61850 and also to perform advanced research in the field.

The core of the substation automation system is the group of protection relay IEDs that perform a variety of functions several relay IEDs have been deployed in the design of the lab to demonstrate all aspects of system protection. The relay IEDs are of differential, distance, and bay controller types from different vendors, and they communicate on different protocols. SA uses various protocols such as Modbus, IEC 60870-103, 101, and currently IEC 61850 for communication. A protocol converter was included in the design of the laboratory to integrate IEDs with different protocols. Because the relay IEDs need to be tested for numerous fault conditions to demonstrate the capabilities and to set the parameters, a secondary test device that generates a variety of fault conditions as per the requirement is an integral part of the design of the lab. As the synchronization of the relay at a single location and also with relays at different locations is done with a GPS clock in the power system, a GPS clock was incorporated into the design of the laboratory to enhance understanding of the actual system. The system architecture of the SA laboratory is given in Figure 4.22.

A superior data highway is a must to transfer the information, and the laboratory incorporates industry standard networking. It has an Ethernet data highway operating at 100 Mbps. The control center designing was intended to meet research and training requirements. Computer systems with appropriate operating systems are used with LCD monitors to perform different functions. A part of the SA laboratory setup is presented in Figure 4.23.

4.11.2   Software components of the SA laboratory

The laboratory utilizes two sets of software programs: hardware-specific and proprietary software and open-ended software that can communicate with any hardware device:

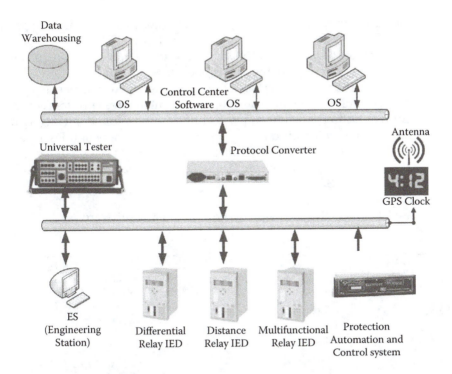

Figure 4.22 (See color insert.) The substation automation laboratory setup.

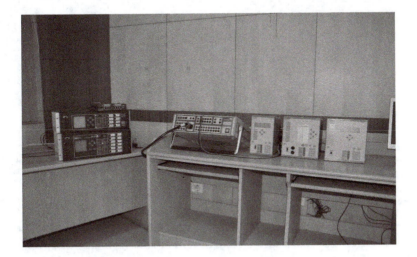

Figure 4.23 (See color insert.) The relay IEDs, secondary test device, and protocol converter.

Proprietary Software: This is dedicated software that is hardware specific; used in the system are Digsi4, ACSELERATOR Quickset, Easy connect, and Omicron software. These programs are used in the laboratory to configure the respective hardware devices.

Control Center Software: The SCADA portal is an open-system software that enables the system to develop highly interactive HMI for remote control. It combines the unique usability features found in HMI with simple integration of control equipment and a variety of IEDs. It can communicate with locally and geographically distributed devices through communication protocols like OPC (OLE Process Control) and MODBUS.

Figure 4.24 depicts the hardware and the associated software.

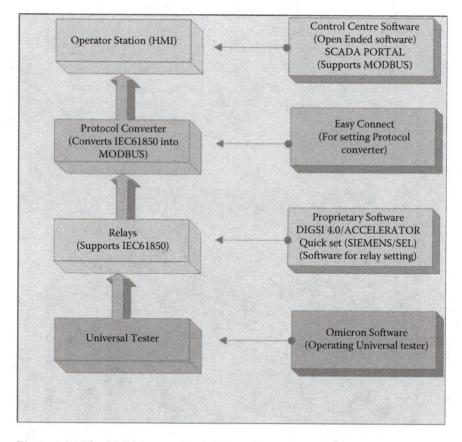

Figure 4.24 The SA laboratory hardware and associated software.

4.11.3 Mitigation from old technology to the new technology

It is important to understand the migration process of moving from old technology to new technology, as most of the old investments may not have completed their operational lives. One such migration option is to employ a protocol converter for IEDs, RTU, and SCADA systems to provide inter-operability between new IEC 61850–based systems and old proprietary or other IEC standard–based systems.

In order to make the students understand this migration process, an attempt has been made to integrate the IEC 61850-compliant IEDs of the laboratory with the available control center software SCADA Portal, supporting Modbus through a protocol converter which involves the following tasks:

- Physical wiring of the devices
- Simulation of field devices
- Integration with control center software
- Configuration of the relays
- Configuration of the protocol converter
- Development of HMI

Thus, all the analog and digital values measured by the relays are displayed correctly at the control center software through the designed HMI, and the retrofitting is thus completed.

The field creation is a formidable task in the SA laboratory experiments. Monitoring and controlling an actual substation were not viable options for obvious reasons at a laboratory level. Hence, the required conditions in the field are simulated and analyzed as per the requirement. In order to simulate different field devices and working environment, the laboratory has software programs PSCAD/EMTDC and MATLAB® and Simulink with Sim Power System Toolbox, by which the field conditions are simulated and played to the relay IEDs using the Omicron secondary test device. The list of experiments in the laboratory includes integration of different relay IEDs, offline and online testing of relay IEDs, integration of protocol converter, GOOSE message testing, and a variety of other innovative experiments as per the course requirements.

4.12 Case studies in substation automation

Substation automation systems are in place across the world in many forms, and the following case studies present a variety of experiences.

Realizing the Power of Data Marts [24]. Utilities may not realize the untapped potential of IEDs installed in the substations as only 20% of the potential benefits are utilized generally. In addition to the already built-in archives or historians at the local control centers, establishment of a corporate data warehouse with operational and nonoperational data can provide a wealth of information to various departments of the utilities. The implementation of such a data mart is discussed here, and the benefits realized by water and power municipal utilities are discussed.

Riverside Initiates Substation Automation, Plans SCADA and Data Warehouse [25]. Riverside Public Utilities (RPU), California, which already had some automation IEDs, required a specific solution to integrate legacy equipment and develop a substation automation system, SCADA, and eventually a data mart. The existing system was upgraded successfully to take care of all the needs of the utility.

ISA Embraces Open Architecture [26]. As early as 1999, the national utility in Colombia, ISA (Interconexion Electrica S.A.E.S.P.), realized the power of standardization by choosing open systems while upgrading and modernizing the substations. This provided more cost-effective monitoring and efficient control of the devices.

Substation Integration Pilot Project [27]. Omaha Public Power district (OPPD) implemented of an automation plan that included EMS, distribution automation, SA and mobile computing, control, metering, and data collection using IEDs. The project included the MMS (manufacturing messaging specification) IED protocol with UCA (Utility Communication Architecture), GOOSE messaging, and generic object models for substation and feeder equipment (GOMSFE).

Plan Ahead for Substation Automation [28]. To match business development to its goals and objectives, Mid-American Energy in Iowa, prepared a substation automation business case for reliability and quality of service, customer loyalty, information to the data warehouse, and cost of service as the main drivers.

4.13   Summary

This chapter provides a description of the substation automation development over the years from RTU-centric architecture to the latest process bus-based digital substation. The islands of automation existing in a substation are discussed, and the integration of functions to make a digital substation based on multifunction IEDs is elaborated. The substation automation application functions are discussed with examples. A complete section is dedicated to data warehousing, and the benefits, problems, and techniques are discussed. The chapter describes the substation automation laboratory set up to demonstrate the system in India and concludes with some case studies from the industry.

Bibliography

1. J. D. McDonald, Substation automation, IED integration and availability of information, *IEEE, Power and Energy Magazine*, vol. 1, no. 2, pp. 22–31, March/April 2003.
2. M. C. Jansen and A. Apostolov, IEC 61850 impact on substation design, IEEE, PES Transmission and Distribution Conference and Exposition, DOI 10.1109/TDC, 2008. 4517219..
3. Klaus-Peter Brand, The standard IEC 61850 as prerequisite for intelligent applications in substations, IEEE PES General Meeting, pp 714–718, vol. 1. Denver, CO, 2004.
4. Mini S. Thomas, Pramod Kumar, and V. K. Chandna, Design, development and commissioning of a supervisory control and data acquisition (SCADA) laboratory for research and training, *IEEE Transactions on Power Systems*, vol. 20, pp. 1582–1588, August 2004.
5. Mini S. Thomas and Anupama Prakash, Design, development and commissioning of a substation automation laboratory to enhance learning, *IEEE Transactions on Education*, vol. 54, no. 2, pp. 286–293, May 2011.
6. Mini S. Thomas, Remote control, *IEEE Power and Energy Magazine*, vol. 8, no. 4, pp. 53–60, July/August 2010.
7. James Northcote-Green and Robert Wilson, *Control and Automation of Electrical Power Distribution Systems*, CRC Press, Boca Raton, FL, 2006.
8. John D. McDonald, *Electric Power Substation Engineering*, 3rd ed., CRC Press, Boca Raton, FL, 2012.
9. M. Kezunovic and T. Popovic, Data warehouse and analysis agent, Fault and Disturbance Conference, Atlanta, GA, May 2008.
10. Brad Tips and Jeff Taft, Cisco smart grid: Substation automation solution for utility operations, white paper.
11. John D. McDonald, SGIP 2.0 and the near future: Our agenda is a window onto likely outcomes in 2014, *Intelligent Utility*, January/February 2014.
12. John D. McDonald, Laying the foundation for a 21st century grid: Interoperability remains the focus for SGIP 2.0, Inc., *Electric Energy T&D's* supplement on "Grid Transformation." January/February 2014.
13. John McDonald, Substation automation and enterprises data management to support smart grid, G.E. Digital Energy, Georgia Tech clean energy speaker series program, March 2011.
14. D. Kreiss, Non-operational data: The untapped value of substation automation, *Utility Automation*, September/October 2003.
15. C. Hor and P. Crossley, Extracting knowledge from substations for decision support, *IEEE Transactions on Power Delivery*, vol. 20, no. 2 , pp. 595–602, April 2005.
16. Ching-Lai Hor and Peter A. Crossley, Unsupervised event extraction within substations using rough classification, *IEEE Transactions on Power Delivery*, vol. 24, no. 4, pp. 1809–1816, 2006.
17. C. L. Hor and P. A. Crossley, Substation event analysis using information from intelligent electronic devices, *Elsevier Electrical Power and Energy Systems*, vol. 28, pp. 374–386, 2006, DOI: 10.1016/j.ijepes. 2005.12.010.

18. J. Jung, C. Liu, and M. Gallanti, Automated fault analysis using intelligent techniques and synchronised sampling, TP 141-0, IEEE PES Tutorial on Artificial Intelligence Applications in Fault Analysis, July 2000.
19. S. Ghosh, Knowledge Extraction in Power Systems, Master's thesis, Department of Electrical Engineering and Electronics, University of Manchester Institute of Science and Technology, Manchester, United Kingdom, December 1999.
20. Mini S. Thomas, Iqbal Ali, and Debajit Nanda, Development of a data warehouse for non-operational data in power utilities, in *Proceedings of POWERCON 2006 Conference*, April 2006. DOI: 10.1109/POWER1.2006.1632499.
21. Mini S. Thomas, D. P. Kothari, and Anupama Prakash, IED models for data generation in a transmission substation, *Proceeding of the IEEE International Conference*: *PEDES*, New Delhi, India, 2010.
22. Mini S. Thomas, Nitin Srivastava, and Anupama Prakash, Waveform extraction and information exchange for data warehousing in power utilities, *Proceeding of the IEEE International Conference: PEDES*, New Delhi, India, 2010.
23. Klaus-Peter Brand, Volker Lohmann, and Wolfgang Wimmer, *Substation Automation Handbook*, Utility Automation Consulting Lohmann, Bonstetten, Switzerland, 2003 (http://www.uac.ch).
24. J. D McDonald, S. Rajagopalan, J. W. Waizenegger, and F. Pardo, Realizing the power of data marts, *IEEE Power and Energy Magazine*, May–June 2007.
25. J. D McDonald, J. Carassco, and C. Wong, Riverside initiates substation automation, plans SCADA and data warehouse, *Electricity Today*, no. 8, 2004.
26. J. D McDonald, D. Caseres, S. Borlase, J. C. Olaya, and M. Janssen, ISA embraces open architecture, *T&D World*, October 1999.
27. T. Nissen and D. Petrchuck, Substation integration pilot project, *IEEE P&E Magazine*, March/April 2003.
28. S. Haacke, S. Border, D. Stevens, and B. Uluski, Plan ahead for Substation Automation, *IEEE P&E Magazine*, March/April 2003.

chapter five

Energy management systems (EMS) for control centers

5.1 Introduction

Electric power systems are the most complex machines in existence, spread over vast expanses, generating electric power at remote locations, transporting it over large distances, and distributing the electricity to consumers across the country. From the time the first high-voltage long-distance transmission line was commissioned in 1917 by American Gas & Electric, there have been tremendous strides in the transmission of electric power. Distances increased between the generation and load centers, voltage and current levels are reaching new dimensions, and the vulnerability of the grid is increasing day by day. The bulk generation and transmission of electric power are still the mainstays of the electricity grid, although *prosumers* with renewable generation at the local level are being encouraged by electric utilities.

The automation of the electric power transmission system spread over wide geographical areas is of utmost importance, and the transmission utilities across the world have embraced the transmission automation in a big way, as it helps the operator in the control room to monitor and manage the system in an efficient manner. The utilities and governments have to make sure that transmission system failures leading to blackouts are avoided and the latest technological advancements are put to use in the control centers to assist the operators in ensuring reliable power.

5.2 Operating states of the power system and sources of grid vulnerability

Power system operation is the balancing act between the ever-changing random loads and the optimized generation so that the cost is at a minimum. A system experiences small disturbances, which may escalate into larger problems, and large disturbances due to sudden large changes, and the operator has to do a balancing act at all times to keep the system stable. The power system designers try to make the system reliable by keeping enough of a margin between the generation and load and additional power transfer capability and actual power flow so that the system has the flexibility to take care of equipment outages for maintenance and sudden failure. However, the additional generation and transmission capacity is limited by the additional investment required; hence, it is the prerogative of the power system operator to manage the system stability within the given margins.

It is worthwhile to understand the operating states of the power system and the attributes for operating the power system in normal state. There are five operating states defined for a power system, depending on the power transfer equation and the equality and inequality constraints of the system. The power transfer equation represents the dynamic behavior of the system with synchronous machines, control mechanisms, and load behavior. The equality constraints represent the total generation that has to equal the total load at any instant of time, whereas inequality constraints refer to the voltages and currents that have to be within certain limits. The identified states are normal, alert, emergency, extremis, and restorative, and Figure 5.1 gives a pictorial representation of the states.

In the normal state, all equality and inequality constraints are satisfied and not violated, and the generation is sufficient to supply the current load and no constraints are violated. When the margin between the generation and load narrows, even a small disturbance can overload equipment and violate the inequality constraints beyond the maximum set limit. This is the alert state, where the equality and inequality constraints are still satisfied. If the system conditions still deteriorate, some of the equipment may get overloaded and the inequality constraints will

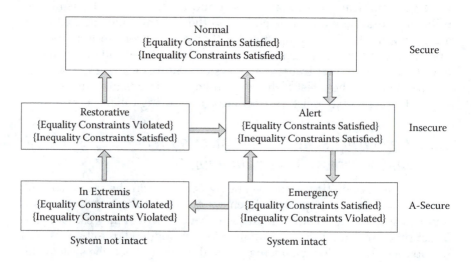

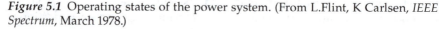

Figure 5.1 Operating states of the power system. (From L.Flint, K Carlsen, *IEEE Spectrum*, March 1978.)

be violated, moving the system to the emergency state. The security margins will be literally zero, and the system will be most vulnerable. Here the equality constraints are still maintained and by immediate corrective action, the system can be brought back to the alert state.

However, if the disturbance continues to escalate or corrective action is not taken at the appropriate time, the system will be plunged into a state of extremis, where all constraints will be violated and the system will lose the load or generation. Under such a scenario, islanding of the system will occur naturally or intentional islanding will be created by the operator. Under these conditions, the operator should be able to take the appropriate action to maintain the islands and avoid complete collapse of the system. Once this is achieved and total collapse is prevented, some of the loads can be reconnected, and generators resynchronized, and the system can then enter the restorative state. Here the demand is not completely met; however, from this state, the system can move to emergency and/or to normal state by appropriate action by the system operator.

Many factors contribute to the change in the states of the system, such as inadequate security margins, failure of protection and control functions and devices, communication system breaks, and so on. Failure of equipment such as transformers or bushings also causes grid vulnerability. Natural calamities are major causes, which cannot be avoided, whereas cyber intrusions and sabotage by intruders can be avoided to some extent by proper firewalls and security systems.

The description of the states of the power system emphasizes the important role the system operator plays in maintaining stability; hence, the energy control center attains significance. From the transmission automation perspective, the supervisory control and data acquisition (SCADA) system delivers valuable system data to the control center, thus informing the operator about the actual state of the system. With the wide-area monitoring system (WAMS) and phasor measurement units (PMUs) adding a new dimension to the measurements, and situational awareness helping to visualize the system better, the energy control centers across the world are changing, and the power system operators are much more aware of the actual status in the field before they take appropriate actions.

5.3   Energy control centers

The energy control centers are the SCADA master stations as discussed in Section 2.8. The power flows in the transmission systems are monitored and managed by the system operators in the centers. The settings of an energy control center thus will be those of a large master station as given in Figure 5.2, where the operator consoles will be manned by system operators with all the required assistance tools. The communication channels bring in data from the field and the set of servers fetches the required data

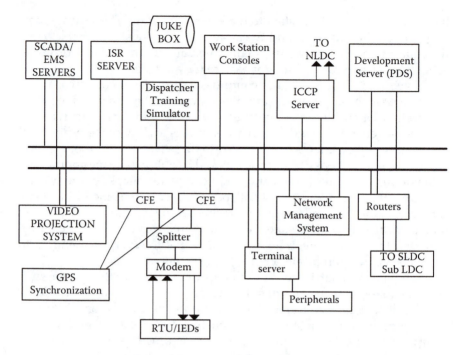

Figure 5.2 Energy control center configuration. (Courtesy POSOCO, PGCIL, India.)

from the master station LAN and displays the appropriate information for the operator to visualize the system. The energy control center will have the basic SCADA software for data acquisition and remote control. The main functionalities of the energy control center, however, are the generation and transmission SCADA application functions or the energy management systems (EMS) functions that are discussed in detail later in this chapter. The energy control center (Figure 5.2) will have a host of server systems including the SCADA server, applications server, information storage and retrieval server, development server, network management server, video projection system, intercontrol center communication server, and layers of security firewalls. The communication front-end processor (CFE) is dedicated to receive data from the field and issue control commands to the field. Detailed discussions of these servers are available in Section 2.8.3.

5.3.1   Energy management systems (EMS): Why and what and challenges [1,2]

In Chapter 2, the basic SCADA functions of data acquisition and control are discussed in Section 2.8.1.2. Also discussed are how these functions are implemented in terms of data acquisition from the field and sending control commands to the field. However, an energy control center also performs a variety of other functions, and collectively these functions are termed *energy management systems.*

Typically an EMS should have

- The knowledge of the complete system to be monitored and controlled, including the parameters of the power system, the interconnections, and so on
- The ability to capture the real-time analog and digital data from the field
- The capability to validate the measured data
- The capability to run the requisite EMS software functions to monitor key system performance indicators
- The capability to send control commands to the field devices and other associated systems
- The capability to display the relevant measured and computed data to assist the operator to make quick and appropriate decisions
- The capability to operate the system within safe limits by tracking the instantaneous load-generation balance
- The awareness of potential risks and the ability to take preventive action
- The ability to start restoration after an emergency in the system or a state change

Thus, the EMS objective is to provide stable, reliable, secure, and optimal power to consumers efficiently and economically. Generally generation and transmission automation systems are termed *SCADA/EMS* systems, wherein the data acquisition and control are SCADA-specific functions.

5.3.2 Energy management systems evolution

With AC transmission systems developing at a fast pace from the beginning of the twentieth century, increasing in size and capacity day by day, the associated reliability, security, and stability of the grid became an issue. The problem was taken seriously after the 1965 blackout of the northern United States which lasted over 13 hours and affected 13 million people. The event led to the creation of two major entities: the North American Electric Reliability Corporation (NERC) and the Electric Power Research Institute (EPRI). NERC developed reliability standards for regions to follow, and EPRI is the centralized research and development organization. Similar organizations came into existence in other parts of the world, and the power engineers started developing standards and strategies to keep the grid stable, secure, and reliable.

The energy control center and EMS also started evolving after the 1965 blackout and have seen tremendous improvements in the recent past due to many reasons, some of which are as follows:

1. Developments in microprocessor technology, making digital computers cheaper and faster with much more computational power
2. Improvements in communication technology making digital communications cheaper and faster
3. Developments in microprocessor-based relays, control equipment, and instrument transformers
4. Global positioning system development and accurate time synchronization including PMU deployment
5. Development of efficient and fast algorithms for EMS software applications, like topology processing, state estimation, contingency analysis, and optimal power flow
6. Situational awareness and visualization techniques providing the dispatcher graphical representation of the power system state for faster, more intuitive understanding

The energy control centers of the 1960s and 1970s were hardwired, used proprietary hardware and software, and worked with analog devices. Analog systems had their own problems, as discussed earlier in the book. The display units were all analog meters and the mimic board was painted, with plugs indicating the opening and closing of switches and circuit breakers that were manually inserted. The limited computations were performed by analog computers. Slowly, with the development of integrated circuits and microprocessors, the control centers also migrated to the digital age. The data acquisition systems developed, and SCADA systems with efficient display and archiving facilities for the modern energy management centers evolved. The system operators with dedicated consoles, modern display and situational awareness tools, intelligent data processing mechanisms, and fast EMS software tools are well equipped to handle the system. (See Figure 5.3 for a depiction of the evolution of the control center from analog to digital.) However, it may be noted that the power transmission business has become more complicated due to the following challenges:

1. Deregulation of the electricity market which presents more challenges, more market-driven approach with independent systems operators (ISOs), regional transmission organizations (RTOs), generation companies (GENCOs), energy traders, and prosumers
2. Aging infrastructure that limits the operating range and increases congestion of the networks

Figure 5.3 (See color insert.) Control center evolution from old analog to new digital.

3. Physical security threats such as terrorism and cyber security threats which challenge the control center operations with unauthorized users
4. Keeping system operators up-to-date with the latest technological advancements in the field to enable them to handle the new technologies and products so that benefits can be reaped by the utility as a whole

5.4   EMS framework

The EMS framework (Figure 5.4) includes the transmission operation management in coordination with generation operation management with the necessary simulation tools and energy services, assisted by the data acquisition and control systems. Every major energy control center will have a dispatcher training simulator where the operators get to analyze past disturbances and create real-life scenarios for study purposes. Figure 5.4 gives a representation of the EMS framework.

The functionalities included in each of the subsystems are

A. Generation operation management
   - Load forecasting (LF)
   - Unit commitment (UC)

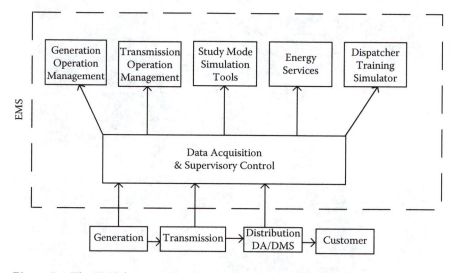

Figure 5.4 The EMS framework.

- Hydrothermal coordination (HTC)
- Real-time economic dispatch and reserve monitoring (ED)
- Real-time automatic generation control (AGC)

B. Transmission operations management: real time
- Network configuration/topology processor (TP)
- State estimation (SE)
- Contingency analysis (CA)
- Optimal power flow and security constrained optimal power flow (OPF, SCOPF)
- Islanding of power systems

C. Study mode simulations
- Power flow (PF)
- Short-circuit analysis (SC)
- Network modeling

D. Energy services and event analysis
- Event analysis
- Energy scheduling and accounting
- Energy service providers

E. Dispatcher training simulator (DTS)

The rest of this chapter is dedicated to the discussion of these functions in detail. It may be noted that the software functionalities discussed here can be categorized in different ways. The above categorization is based on whether the application scenario is transmission related or generation related.

Another way of classifying EMS functionality is to see the time frame in which the simulations are required. The power system operations are set in three time frames, the most critical being the real-time operations, where the operator is assisted by the EMS functionalities to take appropriate control actions. However, a series of functions have to be performed before the real-time operations, to arrive at an appropriate operating plan. Another set of post-event analyses have to be done to calculate the energy transactions, production costs, and so on, and to assess the causes of system contingencies. The above functionalities can be classified in three time frames: real-time operations, pre- and post-real-time analysis. It may be noted that some of the EMS functionalities are performed offline in study mode, as well as in real time to assist the operator (e.g., contingency analysis, power flow analysis, and optimal power flow).

5.4.1   EMS time frames

The EMS functionalities discussed in the previous section can be looked at with a time frame of reference which will give a fair idea about the responses required of each of the software functionalities and also from the system operator.

The RTUs are the eyes and ears of the SCADA system. They acquire data from the field, and the digital data comprising the switch positions, for example, are polled every 2 s. As far as the system stability is concerned, the switch positions are crucial; hence, SCADA systems and telemetry can be considered to have a scan rate around a time frame of 2 s. The automatic generation control that includes the frequency and interchange loops operates around a time frame of 2 s for acquisition of generation, system frequency, and interchange points, and 4 s for issuing generation control commands (raise, lower or set point). The real-time contingency analysis and state estimation runs within a time frame of 60 s to assist the operator with new estimates. Other decision support tools of the EMS such as the topology processor, economic load dispatch, and optimal power flow will run in a 30 min time frame. The dispatcher training simulator is also set up for a time frame of 30 min to train the operator.

Figure 5.5 gives the EMS time frames. The introduction of PMUs in the monitoring scenario creates a time frame in milliseconds (25 to 120 scans per second) and adds a new dimension to the monitoring and control time frame, which will be discussed in later sections.

5.4.2   EMS software applications and data flow

The EMS functionalities run together in a completely coherent manner to move the largest machinery, the power systems, in a secure and stable manner. Figure 5.6 gives the data flow and the major EMS functions in

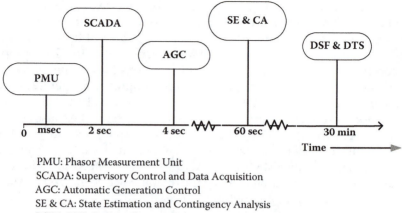

PMU: Phasor Measurement Unit
SCADA: Supervisory Control and Data Acquisition
AGC: Automatic Generation Control
SE & CA: State Estimation and Contingency Analysis
DSF & DTS: Decision Support Functions Dispatcher Training Simulator

Figure 5.5 EMS time frames.

an energy control center. The SCADA system brings in the telemetered system real-time data that are used to create the up-to-date network topology (TP) from the assessment of current switch positions, by modifying the network data from the database. The state estimation (SE) performs observability analysis, detects bad data, adds pseudo data if required, and generates the estimated values of system parameters. The output of SE is used to derive the real-time functions such as power flow, optimal power flow (OPF), security-constrained optimal power flow (SCOPF), contingency analysis (CA), and real-time economic load dispatch (ED).

The following sections will discuss the EMS functions in detail.

5.5 Data acquisition and communication (SCADA systems)

The basic requirement of any automation system is the availability of data from the field, and the SCADA system brings in the required data to the energy control center for further processing and necessary control activity. The control commands are carried back to the field for necessary action. Hence, the SCADA system is the basic requirement for the transmission automation implementation. Communication channels play an indispensable role, as the transmission system is spread over the length and breadth of the landscape. The discussion on basic SCADA functions and the communication of data in Chapters 2 and 3 may be reviewed at this point.

Once the data are available in the control center (SCADA master station), the software applications that help the operator to assess the state of the power system and take appropriate remedial action to keep the stability and security of the system form the basis of the EMS as discussed here.

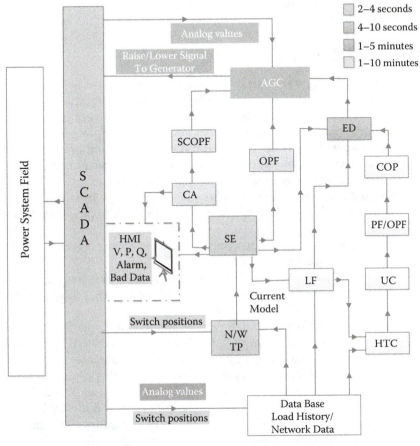

- AGC: Automatic Generation Control
- SCOPF: Security Constraint Optimal
  Power Flow
- N/W TP: Network Topology
- CA: Contingency Analysis
- OPF: Optimal Power Flow
- SE: State Estimator

- ED: Economic Dispatch
- PF: Power Flow
- LF: Load Forecast
- UC: Unit Commitment
- COP: Current operation plan
- HTC: Hydro Thermal Coordination

Figure 5.6 (See color insert.) Energy control center functions and the data flow with time frames.

5.6 Generation operation and management

The proper coordination of the generation operations with transmission operations is the baseline for efficient operation of an energy control center. This section elaborates on the generation operation-related activities and coordination.

5.6.1 Load forecasting [6,7]

The successful operation of a power system depends on the accurate tracking of the system load to match the generation. Predicting the load on the system accurately is the first step in planning of the power system operations. Load forecasting involves many details and requires inputs from the system as well as from the environment and customers. Especially with the widespread renewable sources integration, the accurate prediction of generation is also an issue with solar and wind energy.

The load forecasting needs to be done for various time frames in a system, for smooth control and efficient operation of the power system:

- The automatic generation control (AGC) function ensures that the load-generation balance is maintained online.
- The economic load dispatch function ensures that the larger load variations over a few minutes are distributed among the generators which are available and most economic.
- When the time frames are larger, hours or days, the large load changes are met by the starting up or shutting down of generating plants or importing or exporting power from neighboring areas. The EMS functions used for this purpose include hydro scheduling, unit commitment, hydrothermal coordination, and interchange evaluation.
- The large load variations over weeks are met economically by functions such as hydro scheduling, thermal scheduling, and maintenance scheduling.
- Contingency analysis is utilized to rank the severity of contingencies (offline) and also requires accurate load prediction.

All the above functionalities include accurate prediction of load on the system from the very short term to a few weeks' time, which is generally termed as short-term load forecasting (STLF) in power systems. Medium-term forecasting of electricity demand extends to a 5 year term, and long-term forecasting in electric utilities may extend to 20 to 50 years. The results of these medium- and long-term forecasts are used for planning of generation and transmission with coordination from various departments of the utility and the government. Here we will restrict our discussion to STLF which is used in EMS functions generally.

The state estimation provides corrected data from the measured SCADA data, and this can be used for computing the actual instantaneous load on the system. Very short-term load forecasting techniques are embedded in AGC and in economic load dispatch. STLF provides week-long load predictions to hydrothermal coordination, interchange evaluation, and hydro scheduling. Maintenance scheduling, fuel allocation, and hydro allocation are done for longer periods of 1 to 2 years, and the load predictions for such long periods are obtained from the utility operating plans.

In STLF, the predictions can be classified as

- Short term, where the forecast for the next hour is predicted every 5 min and real-time predictions are made from state estimation data
- Medium term, where the hourly or half-hourly forecast is done for a week and used for daily load planning
- Long term, which is generally done for a year and from which the monthly forecast is done

Electricity demand is closely coupled to the weather of the region, and the inputs to the load forecasting algorithm are invariably the previous load history and the weather data.

The STLF algorithm will have the following inputs:

- The load history is already stored by the SCADA history server.
- The type of day of the forecast, whether it is a working day, a holiday, or a festival day makes a world of difference to the load on the system. It is observed that the load curves for similar days, say every Tuesday, seem to coincide and therefore are cyclic in nature. The past load curve is an effective guide for the current day's prediction.
- Two types of weather data, the weather history of the previous day, week, year, and so on, and the weather forecast for the prediction period are both used.
- The type of algorithm to be used for the forecasting, whether regression technique or stochastic methods, or the intelligent technique; generally, artificial neural network (ANN) algorithms are used as input.
- For longer-term predictions, the data already produced by the system planners needs to be fed to the prediction system.

5.6.2   Unit commitment

Unit commitment refers to committing a generating unit to be online and generating electricity during a specific time slot, so that the generation requirements of the system can be met and other EMS functions can take

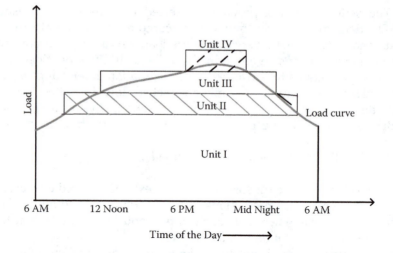

Figure 5.7 Unit commitment of four generating stations.

necessary action. Generating stations such as thermal units require quite an amount of time to come up to the required speed, generate the required voltage, and get synchronized to the system. Once the unit is committed, it should be available to supply load; however, to leave the generating unit committed all the time becomes an expensive affair. It makes much more sense to de-commit the units when they are not required. Once the load forecasting is done, depending on the requirement, it is economical to know when each unit is to be committed and de-committed for specific time slots.

Figure 5.7 illustrates a simple unit commitment example where four units are supplying the load on the system and the units are committed and made available, de-committed, and turned off and when there is a demand as seen from the predicted load curve.

The unit commitment problem is not as simple as discussed above, as each power utility will be working under a large number of constraints and laws, and these have to be built into the unit commitment problem, so that an optimal economical schedule is available for the operations.

Keeping ample spinning reserve as per the directive of the regulatory authority in each country is a constraint. Spinning reserve is the difference between the total generation capacity available of all the units online (spinning) and the load supplied including the losses. It gives an indication of how much extra capacity the system possesses to take care of an emergency, whether the sudden loss of a generating unit or a transmission line. Each utility will have norms for the percentage of generation or load as spinning reserve. Typically, the reserve is equivalent to the largest generation unit, so that the sudden loss of any unit can be handled. The spinning reserve could also be a mix of stations that can be quick started with traditional thermal plants. The spinning reserve can also be geographically distributed so that the transmission corridor congestion may not pose a threat.

Other constraints for unit commitment include the maximum uptime and minimum downtime for thermal units, startup and shutdown time and costs, hydro units that may be in a state where they have to run continuously due to rain and flooding of the reservoir, and other generating units that may have to conserve fuel. Respective utilities may also have their own constraints to be considered while meeting their unit commitment.

The unit commitment defines a specific number of time intervals and the number of units to commit. The permutations and combinations can be exponentially high; however, once the constraints are set and the actual system loading comes into the picture, the number of feasible solutions diminishes. The usual methods used are Lagrange relation, dynamic programming, and priority list schemes that are well established.

5.6.3   Hydrothermal coordination

Hydroelectric generating plants generally co-exist with thermal plants, and the coordination of hydro and thermal plants requires some attention due to the complex problems associated with the scheduling of hydro plants.

Each system that supplies water to a hydro plant is unique in character, and many factors affect the water flow upstream and downstream in a plant. Some plants may have to deal with flood waters, while others may have to release water for irrigation purposes downstream or may be feeding another hydro plant downstream (run of the river plants). Some rivers are used for transport of people and goods and large quantities of water cannot be released at once, and other recreational facilities also may be operational where a water level limit has to be maintained. Each water reservoir may have more than one dam and many tributaries of the river brining in water. The water level in a reservoir is dependent on many factors including weather, rains, and melting of snow.

Hydro scheduling is done in two time frames: long range and short range. Long-range scheduling will start from 1 week and can extend up to 1 year, and will depend on the load, availability of thermal and hydro units, and water flow forecast for reservoirs. The long-range scheduling is to optimize these factors usually using dynamic programming or statistical models. Short-range scheduling is done hour by hour up to a day or maximum one week in advance. Here the load, unit availability, and water inflow are known, and the economic dispatch is done.

Three scenarios can be considered in categorizing the hydrothermal scheduling problem, one in which hydro generation is a small portion of the total, one in which major generation is hydro, and a third scenario in which hydro is the only generation source available.

When hydro systems are only a small component, the scheduling is done to minimize the cost of thermal generation. Generally a majority of the energy control centers experience this scenario as the optimal hydro thermal mix is 30 to 70 for effective and economical scheduling. Systems with only hydro are difficult to schedule and operate, as it is difficult to meet the constraints of all the involved water bodies and the operator does not have a thermal source that can be scheduled without the fear of releasing too much water or disturbing the balance of the surroundings. Systems with a majority of water bodies can be scheduled to achieve a minimum cost of thermal plants.

Hydro scheduling is a complex problem; however, with the proper thermal mix, it can be very effective, economical, and environmentally friendly.

Maintenance scheduling is also a complex procedure, as the equipment in the generation and transmission systems requires routine maintenance and has to be taken out of service once in a while. Maintenance scheduling has to deal with many operating constraints so that the economic impact on the utility is minimal. Once the maintenance scheduling is done, then the unit commitment for a specific period of time can be drawn.

5.6.4   Real-time economic dispatch and reserve monitoring [9]

Once the generating units are "committed" to generate electric power for a specific operating period, it is the duty of the operator to see that the system is operating in the stable state with minimum economic impact. The economic load dispatch solution determines the optimal combination of generating units to supply the load demand subject to the constraints imposed on the system, with minimum operating cost for a specific interval of time, generally one hour.

The simplest economic dispatch problem defines the equality constraint as follows: the sum of the generator outputs should be equal to the load demand, where the losses in the network are neglected and the operating limits are not specifically defined. The power output of each unit should be within the maximum and minimum power generation capacity of the specific unit, which becomes an inequality constraint on the generating unit. The line flow limits on the transmission lines also are included as inequality constraints. The network losses, which contribute a good percentage in some cases, have to be accounted for, which redefines the problem statement to one where the generation must equal the load plus the losses. This involves the computation of the network losses by a suitable method. Hence, economic load dispatch is an optimization problem,

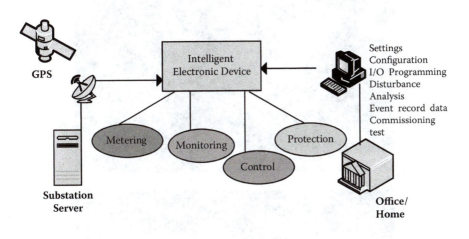

Figure 2.8 Functional view of modern IED.

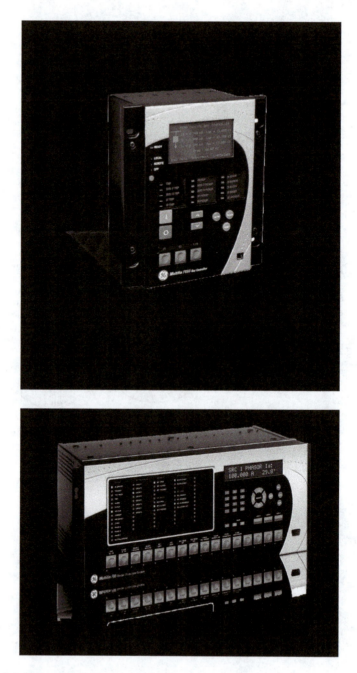

Figure 2.9 Relay IEDs. (Courtesy of GE.)

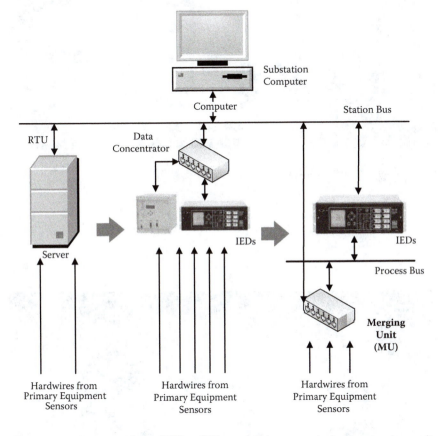

Figure 2.14 Migration from RTU to IEDs and data concentrator to merging units and IEDs.

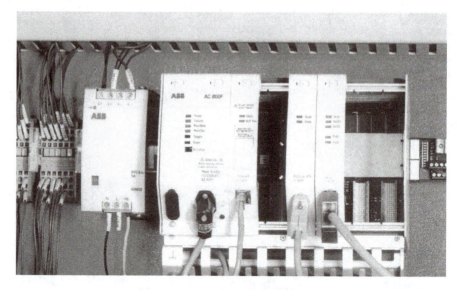

Figure 2.23 Processor of the DPU.

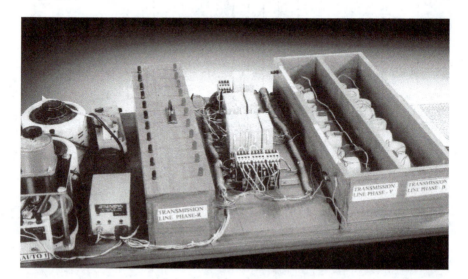

Figure 2.26 Three-phase transmission line model with OLTC, isolators, and transducers.

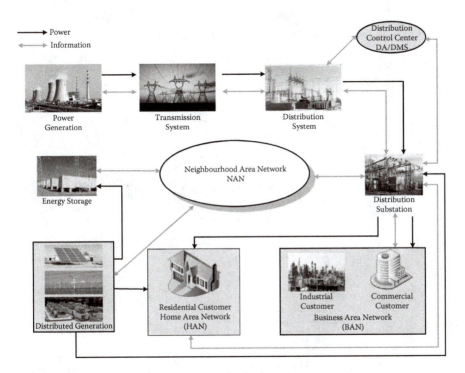

Figure 3.1 Expansion of two-way communication to distribution system including customers.

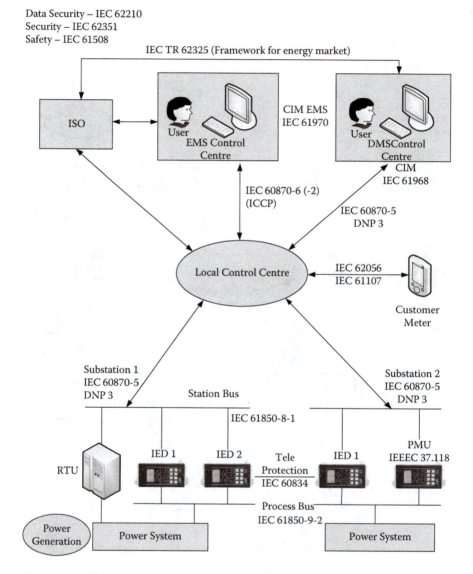

Figure 3.16 SCADA and smart grid protocols in use and under development.

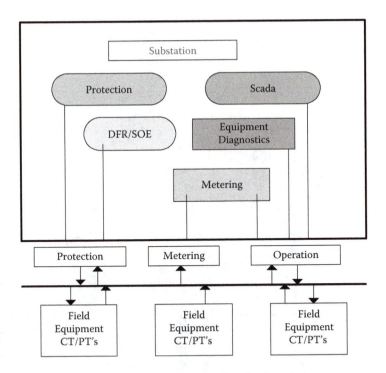

Figure 4.3 Islands of automation in a conventional substation.

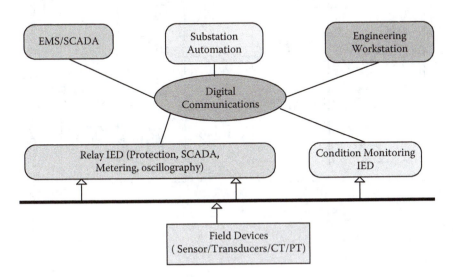

Figure 4.5 Integration of functions in a substation.

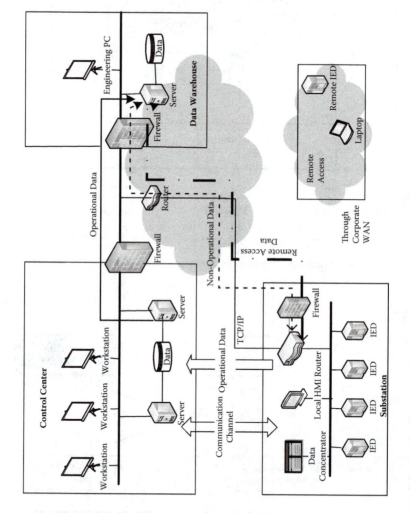

Figure 4.7 Functional data paths (operational, nonoperational, and remote access).

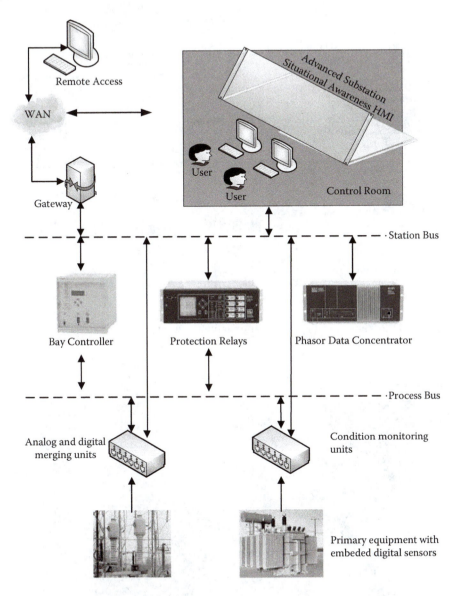

Figure 4.10 The new digital substation.

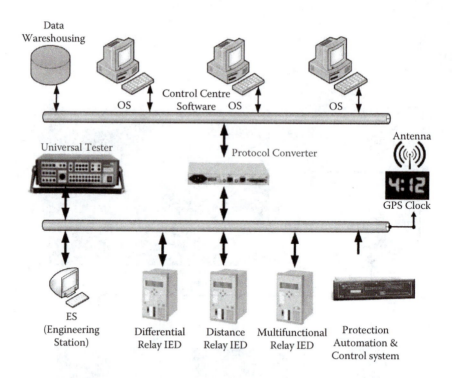

Figure 4.22 The substation automation laboratory setup.

Figure 4.23 The relay IEDs, secondary test device, and protocol converter.

Figure 5.3 Control center evolution from old analog to new digital (will change to POSOCO, India).

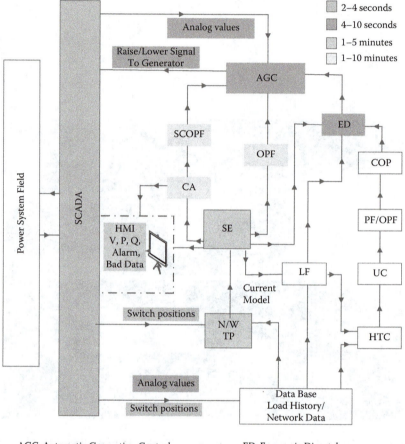

- AGC: Automatic Generation Control
- SCOPF: Security Constraint Optimal
  Power Flow
- N/W TP: Network Topology
- CA: Contingency Analysis
- OPF: Optimal Power Flow
- SE: State Estimator

- ED: Economic Dispatch
- PF: Power Flow
- LF: Load Forecast
- UC: Unit Commitment
- COP: Current operation plan
- HTC: Hydro Thermal Coordination

Figure 5.6 Energy control center functions and the data flow with time frames.

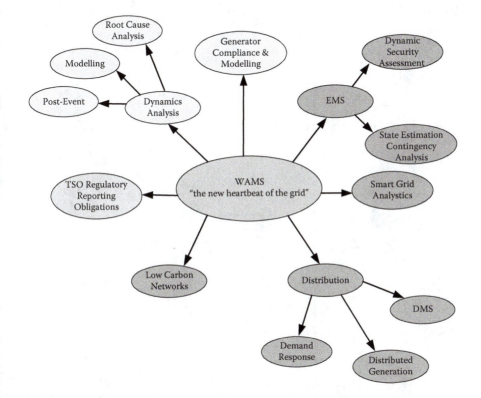

Figure 5.13 Future uses of PMU data in Ems and DMS. (Courtesy Alstom grid)

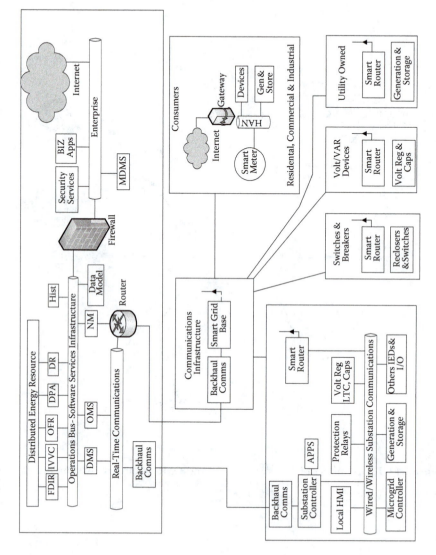

Figure 6.10 DMS integration.

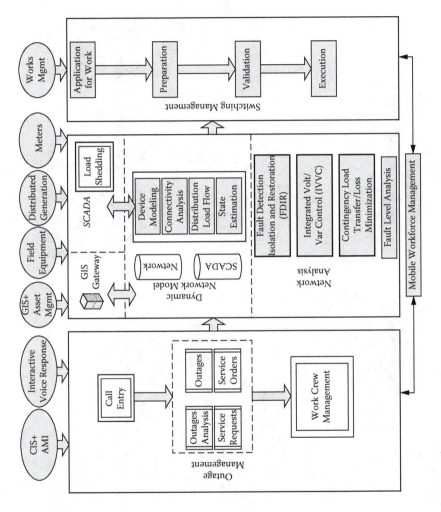

Figure 6.11 DMS framework.

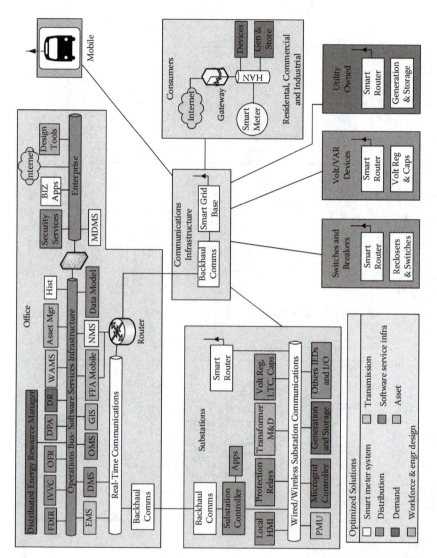

Figure 7.10 Smart grid road map.

where the operating cost is minimized with the equality and inequality constraints satisfied.

There are many solution methods to the economic load dispatch problem, such as Lambda iteration method, dynamic programming, gradient search method, and by using intelligent techniques, such as genetic algorithm, evolutionary computing, particle swarm optimization, and pattern search methods. The result of the economic load dispatch problem gives the optimal power output values (base points) of the constituent committed generating units, which provides minimal overall cost of operation of the units.

The economic load dispatch solution thus provides an operator with the relevant information regarding the generation available and the load served economically. It allows for calculation of the reserve available from among the committed generators and aids in *reserve monitoring*. The aim is to keep the system in the stable state all the time. In case the spinning reserve is depleting and the system is slipping to the emergency state, the operator can alert the generating stations to start new units, if available, or can give instructions to reduce the load by resorting to appropriate measures to bring the system back to the normal state. Reserve monitoring also allows the operator to alert the quick-start generating units like the pumped storage, gas, and diesel units to start and supply power to meet the emergency situation.

5.6.5   Real-time automatic generation control [10]

Automatic generation control (AGC) involves keeping the quality, interchange, and frequency of the generated electric power within the limits, within the framework of the generation-load balance at all times. The voltage generated by the machine is a function of the excitation. The closed-loop control of the excitation loop ensures that the voltage generated by the machine is maintained at the specified value. When it comes to maintaining the frequency of the supply, the problem becomes more global than specific to a particular generator, as the system frequency depends not only on the speed of a specific generator, but on the overall generation-load balance.

The frequency control has two closed loops: the first is the primary control that involves the speed governor of the machine which automatically adjusts the input depending on the speed changes of the generator which is a reflection of the generation-load balance. The secondary frequency control loop which is completed through the power system involves interconnection of other generating units through tie lines. The AGC problem is predominantly a frequency control issue with design of the suitable controllers for controlling the system frequency. The aim is to reduce the area control error (ACE), which is a function of the frequency and the tie line power changes, by applying suitable controllers in the control loop. The ultimate goal is to reduce ACE to zero by raising or lowering the generation so that the load-generation balance is maintained. A wide variety of controllers are available, most of which are model based, and research is occurring to develop applied robust controllers that can handle large power system dynamics and their nonlinearities and uncertainties. The inertia of the power system plays a major role in the frequency control, and the introduction of renewable energy generation poses multiple problems to the system frequency. The tie line power changes should be suitably modified to reflect the renewable generation to have an effective ACE and thus a suitable controller for the AGC.

From a SCADA point of view, the automatic generation control is implemented from a generation control center, where the inputs regarding the system are brought in by the SCADA system. The inputs include the system frequency, power output of each generating unit connected to the system and power flow in each tie line. The AGC program is executed and the outputs that determine the command to be transmitted to each generating unit are conveyed back by the SCADA system for appropriate actuator action.

5.7   Transmission operations and management: Real time

The real-time transmission operations and management functions make use of the inputs from the field by means of the data acquisition system and perform different computations to develop the real picture of the network for the operator.

5.7.1   Network configuration and topology processors

The power network continues to grow, with the number of interconnected buses rising exponentially in the later part of the twentieth century. Over 1000 bus systems came into existence in the late 1980s, but with interconnections and power pools, the network sizes expanded to 8000 in 1995 (American Electric Power, AEP). Further, in 2001, 11,000 bus systems (MISO area) were simulated, which again rose to 32,000 buses in 2004 and further expanded to a million buses at present. So the modeling of the power network is a challenge and due to the large size of the network the software systems of the EMS are expensive and complicated.

The power network is dynamic and the network model also needs updating once the system interconnection changes. For example, when a transmission line goes out of service due to a fault or for maintenance, the component (transmission line in this case) is dead and needs to be removed from the network and a new network created for further analysis. The connectivity matrix also changes due to the tripping of the line. A similar procedure will have to be repeated when a generator is synchronized to the system.

The objective of the network topology processor is to weed out the de-energized or dead system components, establish complete connectivity information, and define the live components in the system. The basic system component data are fed to the topology processor as the primary inputs. The inputs from the SCADA system, such as the measurements of voltage, current, and power flows, and power injection from generating stations, along with the status of switches, are enough to provide a fair picture of the system topology. The topology processor will provide an output of the energized network details and the formation of islands, if any in the system, and the viable islands with generation availability. The topology processor will help display the real live network to the operator with the real-time data superimposed on the network, to inform the operator of the actual status of the network.

However it may be noted that the topology errors have a more significant effect on the rest of the algorithms used by the energy control centers (state estimation, contingency analysis, and power flow) than erroneous network parameters and value measurements, as seen in Figure 5.6.

5.7.2   State estimation

State estimation is the process of assigning value to an unknown state variable in a system based on measurements from the same system as per set criteria. State estimation is applied to many systems where the measured values could be erroneous and will hamper further processing of data and visualization of the system. State estimation was used to accurately predict the positions of aerospace vehicles using the measurements obtained from radar and other inputs which had noise and measurement errors.

In power systems, state estimation is used to provide the best estimate of the values of the variables which are the voltage magnitudes and phase angles of the system buses, from the measurements sent by the SCADA system and available from the network model available.

The values sent in by the SCADA system may have errors due to the following factors:

1. The instrument transformers, which are the primary measuring devices in the field, may have inherent errors and may become saturated.
2. The instrument transformers (CTs and PTs) may be of a poor precision class.

3. The characteristics of these devices will deteriorate with time and environmental factors, especially temperature.
4. The CT and PT outputs are wired to transducers that also may have nonlinearity errors, especially for light load and overload conditions. The transducers may be wired incorrectly.
5. Communication channel surroundings can introduce electromagnetic interference or noise to the measurements which may alter the values. It may be noted that the transducer output for analog values is generally in the 4 to 20 mA range.
6. Loss of data due to communication channel failure poses a serious problem.
7. In power systems there may be many redundant incoming measurements that need to be correlated and filtered before assigning them to the state estimator.
8. A time skew in the sampling of the SCADA field information may occur because all points are not sampled at the same time.

State estimation is the process of assigning values to all the state variables in the system from available measurements and network data. The state estimator will identify and correct anomalies in the data, suppress bad data, and refine the measurements, thus finally giving a set of acceptable state variables, which will be used by the operator as well as for inputs to other computational programs of the EMS. The state estimator thus gives values of voltage magnitudes and phase angles and estimated real and reactive power injections. It also generates an error analysis comparing the measured values and estimated values and a list of bad data detected, which is an advanced application, but of utmost importance. Figure 5.8 gives the data flow in a state estimator.

The steps involved are as follows:

1. The inputs to the state estimator are the measurements and switch status data along with the network topology, already estimated from live components.
2. The observability of the system is first checked, as there may be points from which data are not available due to telemetry failure and other reasons. Computations not observable by the network may lead to erroneous results, and more measurements will make the estimation slower. A balance is to be maintained between the computational speed and the accuracy. Generally 1.5 to 2.8 times the number of state variables to be estimated is an optimum range for redundancy in telemetering.
3. If the measured values are less than the optimum and the system is not observable, then more measurements have to be added to make the system observable. In some cases the points are replaced

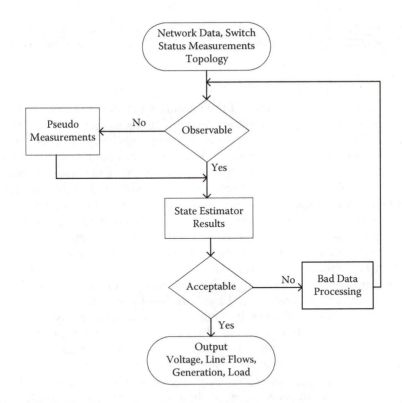

Figure 5.8 State estimator flow chart.

by the previous available value or the expected value, which may be termed as *pseudo measurements*. This measurement is required, as a minimum number of estimated values is required for performing the state estimation successfully.

4. The state estimation is performed, and the results, if acceptable, are used for validating the measurements and other computational purposes.
5. If the results are not acceptable, the data are again processed, bad data are suppressed, the observability is checked, and the process is repeated.

There are many methodologies used for state estimation; the most commonly used ones are the following; however, discussion of these methods is beyond the scope of this book:

1. The maximum likelihood method
2. The weighted least squares method
3. The maximum variance method

The state estimator output values are used as inputs to many programs and operator display by the energy control center as given in Figure 5.6.

5.7.3   Contingency analysis

As discussed earlier, security of the system is of utmost importance, and the SCADA system gives the first warnings to the operator regarding the status of the system. However, power system events happen so fast, in milliseconds and seconds, the operator may not have enough time to perform some kind of analysis with the SCADA data and take necessary action.

Contingency analysis helps an operator model the system contingencies before they happen and see the impact of each contingency on the system like overloading of lines and generators so that appropriate action can be taken when the contingency actually arises. The contingency could be the outage of a transmission line within a control area, or a tie line interconnecting two areas, or a generator outage, to name a few. The inputs to the contingency analysis algorithm are the system parameters and the connectivity, the generation and load on the system, and the voltage set points. The model of the components to be used in the analysis and the rating of each piece of equipment also need to be supplied for the analysis. The analysis will result in the contingency ranking of the component outages with respect to the violations in limits of the lines and generators.

The effect of each of the contingency on the system will be different, the line flows in each line may change, and the voltage profiles may deviate from the prescribed limits. The loss of a generator may result in increase of generation by other units within or outside the control area.

The problem may appear simple, as it involves the solution of the load flow problem with a modified network, as per the outage of each component. However, the large size of the system to be analyzed and the number of contingencies to be studied make the analysis time consuming. In a system with n number of components, contingency analysis for a single component outage will take a considerable amount of time, depending on many factors like the speed of computer systems, the system model in use, and the algorithm. It may be noted that if the contingency analysis takes more time to complete, then the system conditions may have already changed and the analysis needs to be repeated for the new scenario. This necessitates that the contingency algorithm should be fast and should give the ranking quickly to allow the operator to take advantage of the results. As an example, for a 33 bus system, with 40 lines, 18 transformers, 10 generators, and 22 loads, a total of 90 single component outages and additional multiple outages are possible. This necessitates a minimum of 90 AC load flow solutions. With system sizes ranging from a few thousand to tens of thousands of buses, the contingency ranking will take a considerable amount of time.

Hence, the following measures may be taken to reduce the computational time:

1. The initial analysis is performed with reduced system model (equivalents) with very fast algorithms, giving only real power flows (DC load flows).
2. Load flow with less tolerance is also used to identify severe contingencies.
3. Severe contingencies identified are then analyzed with full system model and complete power flow algorithm.
4. Rankings of these severe contingencies are done, usually a performance index or an overload index and voltage index are computed, and the contingencies are ranked.

Contingency analysis gives the operator a second level of system security in addition to the first level information provided by the SCADA data received from the field. The third level of security is provided by the security constrained optimal power flow algorithm.

5.7.4 Security constrained optimal power flow

The economic load dispatch computations in transmission systems do not take into account the contingencies and their effect on the system. The single constraint is that the total generation must equal the load demand plus the line losses. This is similar to the power flow computations done in a transmission system; therefore, in optimal power flow (OPF) the power flow equations are taken as a constraint for the economic load dispatch. We then get the optimal power flow which indicates the power flow in a system at minimum generation cost. The objective function in an optimal power flow can be solved for minimum generation cost or can be modified to show minimum losses in the system. In modern times, with restrictions on emissions, minimum emission criteria can also be taken into consideration. The OPF objective function allows the planning personnel in a control center to see the power flow as per the criteria set by the utility, even for load shedding, for the optimum operational point of view.

Security constrained optimal power flow (SCOPF) is an optimal power flow, performed in a real-time environment with system contingencies taken into consideration, to bring the system back to the normal state from emergency or extremis state. SCOPF allows the system to function within operating limits after a contingency.

To perform SCOPF, the load forecast will give the estimated load values, the SCADA measurements bring in the current values of the system parameters, and the contingency analysis yields a list for particular operating conditions. The power flow is run in real time subject to the constraints, the limits are checked and the control variables are defined. The bus voltages, transformer tap settings, capacitor bank switching, reactive compensation, and load shedding and phase shifting equipment settings, are arrived at after the security constrained power flow is run as it gives an idea about the system parameters under a contingency and helps the operator to get the requisite settings. After the SCOPF is performed, appropriate restoration schemes are evaluated and implemented.

5.7.5   Islanding of power systems

During major power system disturbances, the system may become an island or there may be intentional islanding to save parts of the system. Islands can be of insufficient generation (load rich) and experience a reduction in system frequency, or could be of insufficient load (generation rich) and experience a rise in frequency. The frequency rise or reduction is automatically arrested by low-frequency isolation schemes, load shedding, load rejection, and controlled islanding. However, this task is complicated, as it requires proper coordination between the protection systems, generating stations, and the transmission grid.

In the generation-rich scenario, load rejections will have to be done either by full load rejection where there could be loss of synchronism of the generator and it will pull out and stop, or the main generator breaker will trip. In partial load rejection, the generator is running but part of the load (10 to 30%) is lost.

In the load-rich scenario, to bring the under frequency to normal range, load shedding schemes will be used to match the generation and load. The load is shed in generally 10% steps, with a time delay of six to eight cycles between steps. Load frequency isolation schemes (LFISs) use automatic under frequency relay to isolate one or more generators with matching loads.

5.8   Study-mode simulations

The SCADA control center has to perform a large number of study-mode simulations to arrive at an operating plan for the real-time operation of the power system. These simulations are extremely important for the system as well as for the system planners and operators. These simulations include the following.

5.8.1   Network modeling

Modeling of the power system components is a big challenge because of the complexity of the model to be chosen for each study. There is a need to strike a balance between accuracy of the results and the speed of computation, because the power transmission system is a large network. For example, the transmission line model can be the exact model or a reduced T or pie model. The selection becomes all the more important since some of the computations are to be performed in real time to help the operator make decisions where speed matters.

5.8.2   Power flow analysis

The parameters of importance in a transmission system study are the bus voltage magnitude and phase angle and the real and reactive power flows in the lines. The power flow computations provide these parameters using the appropriate system models. The iteration techniques used include Gauss-Seidel method, Newton-Raphson method, and fast decoupled load flow analysis, to take advantage of the sparse Z matrix of the power system, to name a few.

5.8.3   Short-circuit analysis

Short-circuit studies are generally performed on a power system model for designing the protection system while transmission planning is done. The three-phase fault levels are used to set the short-circuit current interruption capacity of circuit breakers. The fault current levels are also used for coordination and settings of the overcurrent relays. Short-circuit studies thus are considered offline studies.

5.9   Post-event analysis and energy scheduling and accounting

After the real-time monitoring is done, the energy control center has to complete many functions to consolidate the events during the dispatch of power. The main functions performed include the energy scheduling and accounting and post-event analysis.

5.9.1   Energy scheduling and accounting

The transmission control centers have to keep track of the energy production cost and the inter- and intra-area transactions and evaluations during specific periods of time. The energy scheduling and accounting software functionality will have the capability to compute the electricity production cost including the fuel cost and maintenance costs for the units committed. The software will also have interfaces to the independent system operators (ISOs) regarding the scheduled exchanges and actual flows of energy. Various reports have to be prepared for authorities and controlling agencies in different formats regarding power transfer, energy, and a host of user-defined parameters. Displays of the power, energy, and other parameters will have to be provided for clearer understanding. These functionalities are part of the software and are extensively used by the utilities.

5.9.2   Event analysis

After a contingency, the SCADA system would have provided a large amount of data about the event, and a detailed analysis of the data would reveal the causes for the event and yield many insights to help the utility prevent another episode. The event analysis is of utmost significance and has been dealt with in detail in Chapter 4.

5.9.3   Energy service providers [12]

The electricity utilities were vertically integrated with one utility owning the generation, transmission, and distribution of power. Deregulation of the electric industry created horizontal integration which led to the creation of a set of service providers. Generation, transmission, distribution, and ancillary services are now independent entities with competition among entities in the same horizontal segment. Generation companies (GENCOs) compete among themselves to sell the power to the customers. The GENCOs not only generate and sell power but also deal with ancillary services and by-products. Ancillary services are responsible for the transportation of electrical energy under normal and abnormal conditions, the price of electricity being dependent on the fuel prices, fly ash, environmental pollution, and so on. Transmission companies (TRANSCOs) are concerned with transmission of power with reliability, whereas distribution companies (DISCOs) are the buyers of power from the transmission systems, with ancillary and main services. Energy service companies (ESCOs) will coordinate with the customers to get them the best available prices. They allow the customer to select the product and refuse inferior service providers, hence ensuring quality and reliability of power supply. There could be a place for marketers and brokers (BROCOs) to help strike better deals. An energy market authority (EMA) can be formed by governments for future planning and cash flow for better environmental protection and other benefits for sustainable growth.

However, for these players to play in a level field, there is a need for an independent watchdog that will implement the directives of the system regulatory authority to coordinate the market players and to ensure the delivery of quality power to customers. Independent system operators

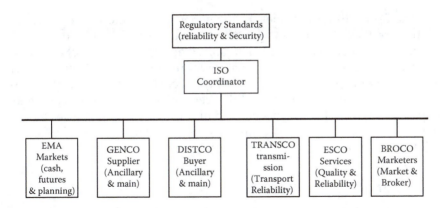

Figure 5.9 Energy markets—the horizontally integrated business model [12].

(ISOs) perform this role and are independent, disassociated agents for market participants and system operations. Figure 5.9 depicts the horizontally integrated energy service market with players controlled by the ISO under the directive of the regulatory authority.

5.10   Dispatcher training simulator

In today's scenario, power system operation is a complex phenomenon, and the security margins are tightened due to the economic and environmental constraints as well as the pressures imposed by the deregulation of the electricity system. The spinning reserves are minimal, and the renewable energy resource integration has put more pressures on the dispatchers to perform miracles. The power system operations have become mission critical, to deliver quality power to the consumers and at the same time to ensure security and safety of the equipment and personnel involved. In addition, the EMS systems are becoming more complex with additional simulations and tools available to assist the operator in the wake of an emergency. All these factors point to the training of operators on systems that mimic the real-time operations where the operators can learn to deal with normal operating situations as well as contingencies.

Dispatcher training simulators (DTS) are installed by the transmission utilities to provide training to the operators, so that they are prepared to deal with the contingencies in real time. The simulator is generally provided by the same vendor who supplies the EMS package to the utility, so that the dispatcher is familiar with the user interface, displays, and control commands.

DTS will have facilities to create real-time scenarios and multiple contingencies to test the trainee, and instructors will assess the actions and responses and provide feedback.

The DTS will have models of the generators, prime movers and the controllers used in the real system, transmission systems, loads, and the relay and control devices. The software applications packages discussed under EMS will be available under the DTS for the trainee to get assistance while performing monitoring and control action during a training session. The instructors build DTS scenarios based on historical events that have happened in the system and also hypothetical scenarios where the disturbances of different kinds are simulated. Experiences of different utilities with the DTS are available in the literature for detailed reading.

5.11 Smart transmission

The transmission systems are automated across the world; however, due to the ever-expanding power demand with the networks getting bigger, the complexity of operation is challenging. The distances between bulk generation and load centers are increasing, and at the same time the challenges of integrating large-scale renewables has added additional constraints to the transmission system. The coordination with generating stations to achieve generation-load balance is the responsibility of the transmission control centers. However, the recent blackouts in many parts of the world have put more focus on the transmission control centers and the way the monitoring and control are implemented. The 2 and 10 s data acquisition by the SCADA systems was not enough for effective monitoring in times of severe contingencies. The phasor measurement units (PMUs) and the broader wide-area monitoring and control systems (WAMS), which allow the transmission network to operate closest to its capacity while maintaining security, have attained significance. The following sections discuss the details of this "smart transmission."

5.11.1 Phasor measurement unit

Synchronized phasor measurements are ideal for monitoring and controlling dynamic performance of the power system, especially under high-stress operating conditions. PMUs were introduced in the 1980s and the wide-area measurement system gathers real-time measurement data across broad geographical areas.

PMU is a digital recorder with synchronizing capability, either a stand-alone unit or part of another protective device like the relay and can time stamp, record, and store phasor measurements of the power system events. According to IEEE standards, IEEE C37.118, a "Phasor Measurement Unit" is "a device that produces synchronized phasor, frequency, and rate of change of frequency (ROCOF) estimates from voltage and/or current signals and a time synchronizing signal."

A synchrophasor is a phasor value obtained from voltage or current waveforms and precisely referenced to a common time base. The PMU extracts the parameter magnitude, phase angle, frequency, and ROCOF from the signals appearing at its input terminals. With better than 1 ms global positioning system (GPS) synchronization accuracy, the PMUs can provide highly synchronized, real-time, and direct measurements of voltage phasors at the installed buses, as well as current phasors of adjacent power branches.

The block diagram of PMU is shown in Figure 5.10. In a PMU, the analog inputs (currents and voltages from the instrument transformers field) are converted to voltages with appropriate shunts or instrument transformers (typically within the range of 10 V) so that they fulfill the requirements of the analog-to-digital converter. The anti-aliasing (low-pass) filter is used to segregate only power frequency signals. The crystal

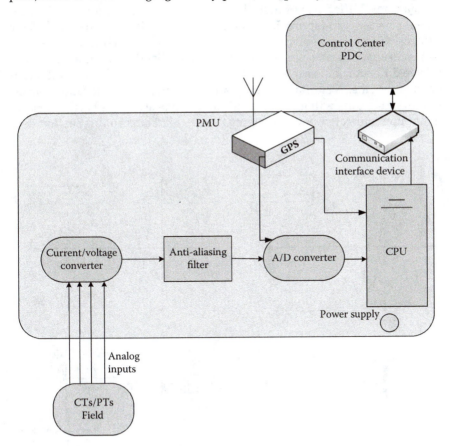

Figure 5.10 Block diagram of PMU.

oscillator in the GPS module provides the sampling clock pulse, which is phase locked with the GPS clock pulse, to the A/D converter, which uses decimation filtering (reduced sampling rate) to convert the analog values to digital values. The CPU module calculates positive-sequence estimates of all the current and voltage signals and stamps it with coordinated universal time supplied by the GPS module. From the CPU module, data are sent through the communication interface module to the control center. The power supply module supplies power to the PMU.

5.11.2 Phasor quantity and time synchronization

Phasor is a quantity with magnitude and phase, with respect to a reference which represents sinusoidal waveform. In a power line connecting bus 1 and bus 2, the voltages V1 and V2 are two sinusoidal waveforms with a phase difference of delta 1 and delta 2 with respect to a reference signal as shown in Figure 5.11. The two voltages, if measured with their respective angular displacements with respect to the reference signal, can be represented as phasor quantities at a centralized location as shown in Figure 5.11, which clearly shows the angular difference between the two ends of the line. A similar representation can be obtained for more buses, and the diagrams will give insights into the phase swings between buses in an interconnected power system, as explained later in this section.

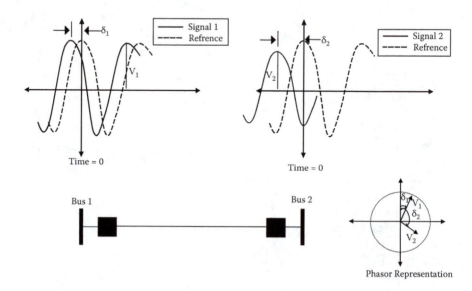

Figure 5.11 Phasor representation of sinusoidal voltages at the ends of a line.

According to IEEE standards, a phasor is defined as "A complex equivalent of a sinusoidal wave quantity such that the complex modulus is the cosine wave amplitude, and the complex angle (in polar form) is the cosine wave phase angle."

The PMU is capable of receiving time from a reliable and accurate source, such as the GPS, that can provide time traceable to coordinated universal time (UTC) with sufficient accuracy to keep the total vector error (TVE), the frequency error (FE), and the rate of change of frequency (ROCOF) error (RFE) within the required limits.

To achieve a common timing reference for the PMU acquisition process, it is essential to have a source of accurate timing signals (i.e., synchronizing source) that may be internal or external to the PMU. For internal, the synchronization source is integrated (built-in) into the PMU (external GPS antenna still required), and for external, the timing signal is provided to the PMU by means of an external source, which may be local or global, and a distribution infrastructure (based on broadcast or direct connections).

Within a PMU, a phase-locked oscillator is used to generate the time tags within the second. The time tag is sent out with the phasors. If a phasor information packet arrives out of order to a PDC (phasor data concentrator), the phasor time response can still be assembled correctly; however, if the GPS pulse is not received for a while, the time tagging error may result in significant phase error.

5.11.3   PMU-PDC system architecture

The synchrophasor standard and associated communication protocol were designed to aggregate data from multiple locations. As each dataset is transmitted synchronous to the top of the second and as each transmitted dataset contains a precise absolute time stamp, the data aggregation function becomes a simple matter of combining sets of data with common time stamps.

PMU architecture consists of PMUs at substation level and PDCs at control center level as shown in Figure 5.12.

A PDC collects PMU data, time aligns the data, and sends it to a computer or another PDC. It forms a node in a system where phasor data from a number of PMUs or PDCs are correlated and fed as a single stream to other applications. The PDCs correlate phasor data by time tagging to create a system-wide measurement set. A PMU at the substation level collects the estimated measurements and feeds them to local operator level or PDC at the control center level. PDC makes decisions with a very high speed of 10 to 100 ms. PDCs are connected to the energy control center and integrated with SCADA for monitoring, protection, and control applications.

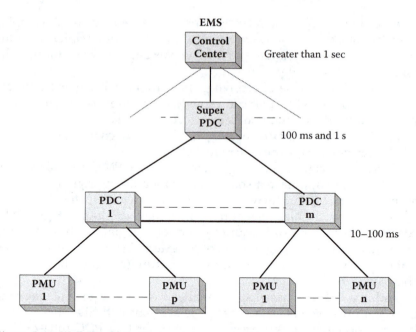

Figure 5.12 WAMS architecture.

5.11.4 Applications of PMU

Initially, when the first commercial PMUs became available, the only application was post-event analysis as PMUs were expensive and the communication channels were not available for real-time data transmission.

Now with the large number of PMUs installed around the globe rapidly, it is worthwhile to look at the applications of phasor data for real-time monitoring, control, and protection applications.

Power System Monitoring: State Estimation. Existing state estimation algorithms use SCADA measurements to estimate all bus voltage magnitudes and angles. The state was inferred from the unsynchronized power flow measurements using a nonlinear state estimator, as there was no provision to measure the state directly. There are many factors that create error here, such as the time synchronization of the measurements, the assumption that between scans, the state did not change and remained static, and so on. This assumption of a static state between the scan intervals could be completely wrong as the state would have changed. With PMUs the scans are faster, the angular measurements are directly transferred to the energy control center and then computed, and hence power system monitoring with PMUs will be more accurate and desirable.

State estimation with only PMU data will require a large number of PMU installations across a network so that the network is completely observable and there are enough communication facilities to process data in a timely fashion. The PMU data will be received with accurate GPS time tagging and also the state estimation problem will have a linear solution with no iterations. PMU measurements include line currents which makes it easier to compute voltage measurements at the other end of the line from the measured end. Optimal placement of PMUs has attracted a lot of attention, and a large number of methods have been proposed in the literature.

Power System Control. Generally in power systems, local control is prevalent with the generators controlled with local measurements and a model of the rest of the power system. However, with PMU data available from a larger system, it is possible to have control with global values at hand. The phasor measurements are time tagged, and the fast data streaming allows the system to have control based on the actual state. The transient stability, electromechanical oscillations, and overload may be identified and corrected on a system-wide basis with PMU data.

With PMU data, a combination of local and global control algorithms will help bring the system back to a normal state in case of an emergency. The angular separations of the voltage phasors may be monitored to detect any instabilities in the system, and appropriate identification of the area of origin can help apply local control to mitigate the problem. The local controllers will have local input signals and will have additional signals from the system-wide PMU-based measurement system.

The latency involved with data transfer may pose a problem for control applications; however, many of the oscillations are in the range of 0.2 to 2 Hz and hence may be detected on time and appropriate action can be taken. PMU will be an ideal means to track the dynamic system behavior in real time and take control action to prevent the system from deterioration and damage.

Power System Protection. WAMS offer a wide variety of solutions to protection problems in power systems. In general, phasor measurements are particularly effective in improving protection functions, which have relatively slow response times. For such protection functions, the latency of communicating information from remote sites is not a significant issue. Some functions that can be better implemented are backup relay protection, adaptive out-of-step protection, and prevention of false tripping of relays in case of a highly loaded system.

5.11.5   WAMS (wide-area monitoring system)

Wide-area monitoring systems (WAMS) are essentially based on the new data acquisition technology of phasor measurement and allow monitoring transmission system conditions over large areas to detect and counteract grid instabilities. Current, voltage, and frequency measurements are taken by PMUs at selected locations in the power system and stored in a data concentrator every 100 ms. The measured quantities include both magnitudes and phase angles and are time-synchronized via GPS receivers with an accuracy of 1 ms. The phasors measured at the same instant provide snapshots of the status of the monitored nodes. By comparing the snapshots with each other, not only the steady state but also the dynamic state of critical nodes in transmission and subtransmission networks can be observed. Thereby, a dynamic monitoring of critical nodes in power systems is achieved. This early warning system contributes to increased system reliability by avoiding the spreading of large area disturbances and optimizing the use of assets.

Most programs for WAMS technology have three stages of implementation. The first stage is to install PMU and PDC, and collect and archive phasor data from important locations throughout the grid to determine the topology and operating limits. In the second stage, the data gathered along with real-time phasor and frequency measurements are used to calculate grid conditions using analytical functions to make suggestions to the grid operator to keep the grid stable and reliable. The final stage is to carry out all of the above functions automatically.

5.12 EMS with WAMS

Wide-area monitoring systems will revolutionize the EMS functionalities as PMU data will provide improved visibility to the grid conditions and hence can provide earlier warnings than SCADA data.

Table 5.1 gives a comparison of SCADA data and PMU data.

Table 5.1 Comparison of SCADA Data and PMU Data

SCADA data	PMU data
Scan rate: 2 s	Scan rate: 25–30 samples/s
Magnitudes of voltage, current, and frequency from the field	Angular difference between measured values from the filed
Latency in the measurements due to the existing old communication infrastructure	Latency is minimal due to the new communication technologies
Not fast enough to respond to the dynamic behavior of power systems	Fast enough to depict the system dynamic behavior
Time stamping for specific values and instances	Completely time tagged data with GPS synchronization

The PMU data will improve security of the system as the data can warn the operator of any instability in the system. The EMS functionalities discussed in the earlier sections described the analysis carried out using system models where as in the new regime with PMU implemented, the computations will be based on the more accurate PMU measurements.

The benefits the PMU measurements bring in can be summarized as below:

1. *State Estimator Performance*: The voltage and current phasors from the PMU can be used in the linear state estimator, and the nonlinear state estimation can be completely eliminated if enough observable PMU data are available. Once the state estimation is accurate and fast, all the other EMS functionalities like contingency analysis, SCOPF, economic load dispatch, and finally AGC will be more accurate and faster.

2. *Improved Stability Margins*: PMU can be used to monitor small signal oscillations in a system and take corrective actions, thus improving transient stability and voltage stability margins.

3. *Disturbance Location*: Using the PMU data and SCADA information, it is easy to locate the origin of a disturbance before it escalates into a major problem. This is done using the PMU measurements which can detect the traveling wave generated by the sudden disturbance.

4. *Postdisturbance Recovery*: PMU can help in the islanding recovery in case of an islanding or for a black start of the system. With PMUs, the islanding is detected fast and an alarm can be raised. The situational awareness will clearly show the island to the operator, and the restoration can be initiated quickly. In case of a black start, the system will be clearly visible to the operator for better coordination.

5. *Disturbance Analysis*: The accurately time stamped PMU data can easily recreate the sequence of events in case of a disturbance for post fault diagnostics.

6. *Situational Awareness*: Improved visualization with the PMU data which indicates the power swings and oscillations area wise, that will help the operator to detect system disturbances from the inception. This will help the operator to make better assessment of the system and make quick decisions during an emergency, and any further oscillations, tripping, and blackouts can be avoided.

7. *Improved Protection Schemes*: With the PMU measurements, some relaying functions could be improved in a highly stressed out system, as explained earlier. Adaptive relaying can be easily implemented, as the real-time power measurements from the ends of transmission lines will be available at the control center by means of PMU measurements.

5.13   Future trends in EMS and DMS with WAMS

PMU applications will deliver instant data to the control center and help the operator to take appropriate action. Thousands of PMUs are being installed across the globe by various utilities; however, consolidating the data, analyzing, visualizing, and putting it to effective use are challenges. It is going to take a long time to integrate the SCADA and PMU data for better coordination and situational awareness implementation.

All the above discussion was based on the PMU used as a measuring device, but the control capabilities of the PMUs are yet to be harnessed, and research work on such implementations continue.

PMU application is not restricted to transmission systems; it has innumerable applications in the distribution sector as well which include demand response, renewable generation, and integration with distribution management systems (DMSs) as shown in Figure 5.13, as PMU becomes the new heartbeat of the power grid.

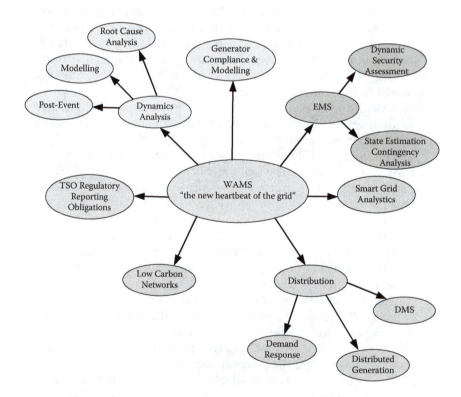

Figure 5.13 (See color insert.) Future uses of PMU data in Ems and DMS. (Courtesy Alstom Grid.)

5.14 Case studies in EMS and WAMS

SCADA and EMS installations have been implemented all over the world to monitor and control the generation and transmission systems and to maintain system stability by coordinating the generation and transmission of electrical energy. With the WAMS improving the visualization and situational awareness in the control centers, consider the example of a pilot project commissioning PMUs in the Northern region of India [21]. This demonstrates the big plans for the fast-growing electricity sector in India.

The pilot project implementation focused on locating and implementing four PMUs, GPS synchronization, integration of data to a central PDC, and visualization of the data to the operator in the control room. The challenges of utilization of the PMU data to a central PDC and integration with the existing SCADA system are discussed in the reference. This pilot project is being extended to the unified real-time dynamic state estimation and measurement (URTDSM) project with 1732 PMUs and 32 PDCs.

5.15 Summary

This chapter discusses the extended SCADA applications to transmission systems, which are termed EMS. The SCADA brings in data from the field, which are used by the energy control centers to perform a large number of application functions in combination with the network models and other system data. The chapter begins with a discussion on operating states of the power system and the evolution of control centers, and the subsequent sections discuss the EMS framework. The EMS functionalities in generation operation and management, transmission operation and management, and study mode simulations are discussed in detail. The dispatcher training simulator and the post-event analysis functions are also explained. The later part of the chapter focuses on PMU and WAMS and compares EMS with WAMS.

Bibliography

1. J. Giri. Enhanced power grid operations with a wide-area synchrophasor measurement and communication network, IEEE Power and Energy Society General Meeting, 2012.
2. J, iri, M. Parashar, J. Trehern, and V. Madani, The situation room: Control center analytics for enhanced situational awareness, *Power and Energy Magazine, IEEE*, vol. 10, no. 5, pp. 24–39, 2012.
3. E. Vaahedi, *Practical Power System Operation*, Wiley-IEEE Press, New York, 2014.
4. Lester H. Fink and Kjell Carlsen, Operating under stress and strain, *IEEE Spectrum*, March 1978.

5. S. K. Soonee, Vineeta Agarwal, Anamika Sharma, and Akhil Singhal, State estimator at National Load Despatch Centre, India, Implementation Experiences, *GRIDTECH 2012*, New Delhi, India.
6. Tao Hong, Jason Wilson, and Jingrui Xie, Long term probabilistic load forecasting and normalization with hourly information, *IEEE Transactions on Smart Grid*, vol. 5, no. 1, pp. 456–462, January 2014.
7. George Gross and Franciso Guliana, Short term load forecasting, *Proceedings of the IEEE*, vol. 75, no. 12, pp. 1558–1572, December 1987.
8. Sasan Mokhtari, Jagjit Singh, and Bruce Wollenberg, A unit commitment expert system, *IEEE Transactions on Power Systems*, vol. 3, no. 1, pp. 272–277, February 1988.
9. Leandro dos Santos Coelho and Chu-Sheng Lee, Solving economic load dispatch problems in power systems using chaotic and Gaussian particle swarm optimization approaches, *Electric Power and Energy Systems*, vol. 30, pp. 297–307, 2008.
10. Hassan Bevrani, Fatemeh Daneshfar, and Takashi Hiyama, A new intelligent agent-based AGC design with real-time application, *IEEE Transactions on Systems, Man, and Cybernetics—Part C: Applications and Reviews*, vol. 42, no. 6, pp. 994–1003, November 2012.
11. Sebastien Gissinger, J. Philippe Chaumes, Jean-Paul Antoine, Andri Bihain, and Marc Stubbe, Advanced dispatcher training simulator, *IEEE Computer Applications in Power*, pp. 25–30, April 2005.
12. Allen J. Wood and Bruce F. Wollenberg, *Power Generation, Operation and Control*, 3rd ed., Wiley, New York, 1996.
13. Gerald B. Sheble, Energy service providers, *IEEE Power and Energy Magazine*, pp. 12–15, November–December 2003.
14. J. De La Ree, V. Centeno, J. S. Thorp, and A. G. Phadke, Synchronized phasor measurement applications in power systems, *IEEE Transactions on Smart Grid*, vol. 1, no.1, pp. 20–27, 2010.
15. Debomita Ghosh, T. Ghose, and D. K. Mohanta, Communication feasibility analysis for smart grid with phasor measurement units, *IEEE Transactions on Industrial Informatics*, vol. 9, no. 3, pp. 1486–1496, August 2013.
16. Ali Abur and Antonio Gomez, Power System State Estimation–Theory and Implementation, Exposition, 2004.
17. S. A. Soman, S. A. Khaparde, and Subha Pandit, *Computational Methods for Large Sparse Power Systems Analysis: An Object Oriented Approach*, Kluwer Academic, Amsterdam, Netherlands, 2002.
18. Mini S. Thomas, Anupama Prakash, and Nizamuddin, Modelling and testing of protection relay IEDs, *IEEE International Conference (POWERCON)*, New Dehli, DOI 10.1109/ICPST. pp. 4745389, 2008.
19. Parmod Kumar, V. K. Chandna, and Mini S. Thomas, Ergonomics in control centre design for power system, *IEEE International Conference, PIC*, New Delhi, India, 2006.
20. Mike Adibi, *Power System Restoration: Methodologies and Implementation Strategies*, Wiley-IEEE Press, New York, 2000.
21. V. K. Agrawal, P. K. Agarwal, and Rajesh Kumar, Experience of commissioning of PMUs pilot project in the Northern region of India, National Power System Conference (NPSE), 2010, Hyderabad, India.

chapter six

Distribution automation and distribution management (DA/DMS) systems

6.1   Overview of distribution systems

Distribution systems are mostly below 69 kV level and supply all their customers with electricity, whether it is bulk supply at higher voltage levels or residential service at the lowest voltage levels.

The distribution system starts from the substation, which hosts the substation transformer, from where the primary feeders emerge and provide electricity to the bulk customers and also to the secondary distribution system. The distribution transformers step down the voltage further, and the distributors carry the power to the individual commercial and household customers. Distribution systems are largely radial and have single, two-, or three-phase lines. Distribution systems are responsible for delivering power to the customers and have to make sure that the customer interruptions are reduced to a minimum and quality power is supplied at minimum cost. Figure 6.1 provides a sketch of a typical distribution system.

6.2   Introduction to distribution automation

All over the world, distribution automation has taken center stage and is the focal point of most of the developments owing to the tremendous opportunities ahead. Smart grid implementations with customer participation which necessitate two-way communications along with renewable integration are forcing utilities to automate the distribution systems in a big way. Automating the substations, feeders and customers is essential for complete automation of distribution systems. Distribution automation (DA) is a set of technologies that enables a utility to remotely monitor, coordinate, and operate distribution components in a real-time mode from remote locations (IEEE PES Distribution Management Tutorial, January 1998).

Distribution automation is all about automating the entire distribution system operations. It will include the basic automation functions like

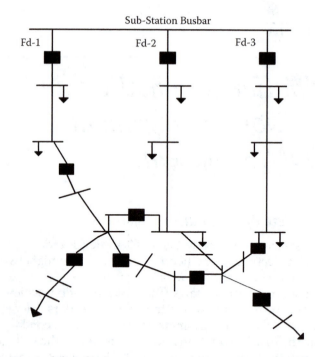

Figure 6.1 A typical distribution system.

supervisory control and data acquisition (SCADA) with all the communication requirements, automating the protection system to outage management systems including customer information systems, and managing assets and data warehousing.

Further, distribution automation encompasses two terms used by industry which are distribution management systems (DMS) and distribution automation (DA) systems (Figure 6.2).

DMSs are associated with operator-focused activities that happen in a control center using the real-time data from SCADA systems and the information from the manually operated devices in the system along with the customer information.

Distribution automation systems are part of the DMS and include all SCADA-related remote monitoring and control actions. This includes the automation spread over entire distribution systems from substation to feeder to customer and the communication infrastructure for implementation of the functions. Distribution automation systems provide real-time monitoring and control functionalities for the downstream network.

Hence, in a distribution utility there are three areas where automation can be implemented: first at the customer location, second on the feeders, and finally in the substation; each has pros and cons, but substations are the typical focal points because they are the greatest source of

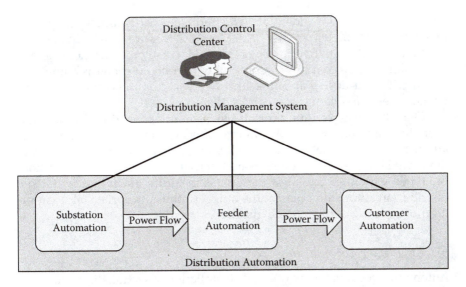

Figure 6.2 Distribution automation and distribution management systems (DA and DMSs).

information. The automation of substations is essential, and utilities all over the world are automating substations.

Every utility has varying needs for automation, depending on the age of its infrastructure, service area demographics, and degree of existing automation. For this reason, all three sources of distribution automation are considered in developing a strategic implementation plan for automation. The following sections give an overview of automation, feeder automation and substation automation.

6.2.1   Customer automation

Customer automation focuses on the smart meters installed at the customer premises. Automatic meter reading (AMR) technology and services associated with it, such as automatic connect and reconnect, are gaining in popularity. AMR periodically records meter readings and relays this information back to a utility office. Installed at the customer location, AMR replaces a human meter reader and is the focal point for home automation implementation, demand response, and associated smart grid functionalities.

Implementing customer automation is cost intensive, as the expenses include the meter and the communications link, usually a wireline (telephone) or wireless connection that transmits data to the utility office from each individual user location. For large commercial and industrial customers that use a lot of power, this expense is quickly repaid. But for tens of thousands of residential customers, the cost is difficult to justify to obtain simple meter reading data. More services are being added to the AMR, and with initiation of the smart grid greatly affecting customers, establishing a live communication link with each customer is a must for the distribution utility, and that link will serve as the conduit through which other valuable services can be offered.

Once two-way communication has been established to a residential user, the utility has opened the door to offering such lucrative services as all the smart grid–related functionality discussed in Chapter 7, remotely monitored home security, high-speed Internet access, and cable television. But the key to providing these services is including them in the strategic distribution automation plan at the outset so the appropriate telecommunications infrastructure can be installed.

6.2.2   Feeder automation

Automating feeders typically entails installation of sectionalizing devices, or switches, along the feeder. When there is a problem with the feeder, data will be fed back to the substation or control center for analysis. Once the problem has been identified, a technician can remotely activate the switch to isolate the segment causing the trouble and reroute service to sections on either side of the problem, or this process may be done automatically. Many other services related to the quality of power supplied are implemented by automating the feeder, like voltage improvement to manage load and capacitor placement and reactive power control to reduce losses.

The challenge of feeder automation is similar to that of AMR—feeders are numerous and are spread over large geographic areas, making installation and maintenance of two-way communication an expensive proposition. As a result, feeder automation is often limited to the 10 or 15 worst performing feeders. By concentrating on problematic feeders, utilities spend less money and can guarantee their automation investment will pay off in reduced duration and frequency of outages. This approach for targeted feeder automation is likely to remain standard procedure in distribution automation projects.

Feeder automation will revolve around monitoring and controlling the feeder sections using the SCADA components discussed in Chapter 2. The feeder remote terminal units (RTUs) and remote switches with communication facility installed on the poles will send analog and status data to the substation via a suitable communication facility. The master station functionality may be at the control center (centralized architecture) or at the substation (distributed architecture) where the automation functions can be performed, and the commands are communicated back to the respective devices on the feeder for action. In Figure 6.3 the equipment is installed at the general office, which includes a master station server,

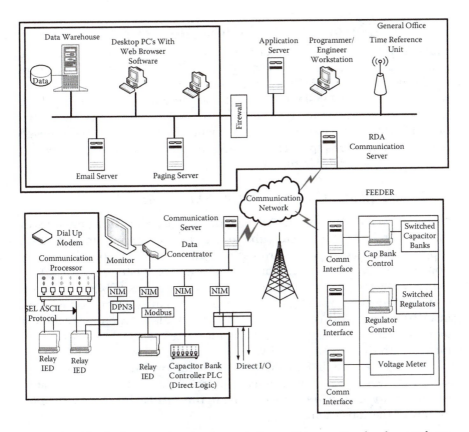

Figure 6.3 Feeder automation implementation with capacitor bank, regulator, and voltage control.

engineering workstation, feeder communications server, and facilities for connecting these devices to the substation communication network as well as the corporate network. One unique feature of this master station is that the system does not include any dispatcher workstations; utility personnel access the system using Web browser software and "remote desktop" facilities running on existing desktop PCs.

A typical initial step in implementing feeder automation is to install a tie switch on a feeder between two substations. Often referred to as using half switches, the method provides the capability to shift feeder load segments from one substation to another.

6.2.3   Substation automation

Substation automation also involves the functionalities associated with the system and has been dealt with separately in Chapter 4.

6.3   Subsystems in a distribution control center

SCADA systems monitor and control the downstream devices and circuits in a distribution substation or control center. The essential components of a SCADA system, the RTUs or intelligent electronic devices (IEDs), transmit data to the master station through the communication media, where the operators monitor and control the operations of the system. The distribution SCADA system has the basic functions with many superimposed application functions specific to distribution management systems, which will be dealt with later in this chapter.

6.3.1   Distribution management systems (DMSs)

Distribution management systems include the real-time functionalities of distribution SCADA coupled with the relevant application functions with support from the corporate process systems such as customer information systems (CISs) and geographical information systems (GISs). DMSs are also integrated with outage management systems (OMSs) and asset management systems (AMSs). In the present scenario, advanced metering infrastructure (AMI) is an integral part of any distribution management planning and discussion, and AMI is integrated with DMS for common information sharing and activity.

The subsystems integrated with DMS for providing quality power supply to the customers with maximum reliability with minimal cost to the utility are shown in Figure 6.4. The figure clearly shows the data integration of the subsystems.

6.3.2   Outage management systems (OMS)

An outage management system is a critical subsystem, where the distribution network is brought back from a state of emergency to normal state, in a minimum time frame, with disturbance to the least number of customers. Outages are sustained interruptions in the power supply to the customers. OMS includes functions such as trouble call management, outage analysis, crew management, and reliability reporting. Outages can be classified as unplanned and planned.

6.3.2.1   Unplanned outages

Outages in a distribution system can occur when a fuse or recloser or a circuit breaker operates to clear a fault and the customers located downstream lose power. This may be due to the sudden failure of a component such as transformers, insulators, and so on. The information about the failure is available to the control center via trouble calls from the customers and switch status changes from SCADA, and also the maintenance crew may detect the fault or outage.

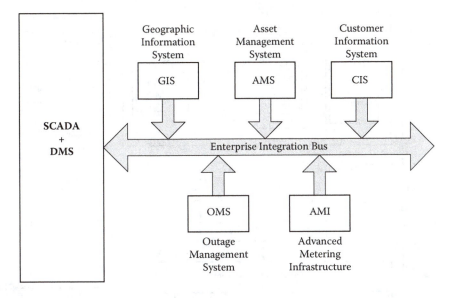

Figure 6.4 SCADA plus DMS integration with other subsystems in a distribution control center.

The OMS will work differently on systems, depending on the level of automation of the distribution system and also by the number of customers served by a distribution transformer. In automated systems, the outage of a component will be known to the SCADA system before any trouble call comes from customers. Especially with an automated metering infrastructure in place, the outage event will be reported to the DMS within a matter of seconds. The OMS program can continuously process and analyze incoming SCADA messages and the trouble calls to locate the fault or outage and the loss of power to customers. The system can also work out the time required to clear the contingency and inform the customers accordingly. The work management and related crew management can be initiated from the OMS as shown in Figure 6.5. Interactive voice response (IVR) systems generally permit trouble call entry into the OMS without human intervention, and the OMS can inform the customers about the outage status already inferred from the SCADA AMI, provide a restoration schedule, and also call back the customers later to verify the supply availability.

6.3.2.2 *Planned outage*

Planned outages are scheduled by the utility for routine maintenance or replacement of equipment. Customers are generally informed in advance about these outages. Planned outage can also be due to the load

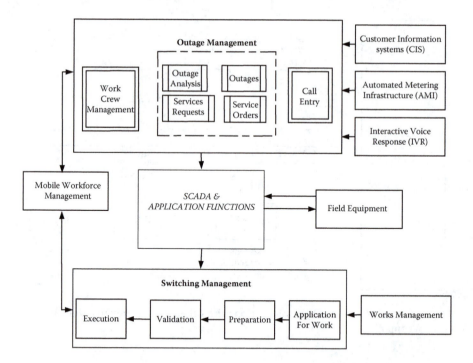

Figure 6.5 Outage management systems.

management algorithm implemented by the operator to maintain the load within the incoming supply limits.

Planned outages are also handled by the OMS and crew management, and informing the customers in advance has to be completed before the planned outage.

6.3.3 CIS (customer information system)

Generally, a CIS is the interface of the utility with the customer. A CIS will store the following information to be accessed by customers at any time:

- Customer data (e.g., customer demographic, customer type and category)
- Meter (e.g., meter type and data)
- Payment (e.g., payment pattern, payment history, payment method, billing history, and billing information)
- KWh consumption (e.g., consumption history and pattern)
- Rate (e.g., rate category and price)
- Irregularities (e.g., any irregularities in payment and consumption patterns)

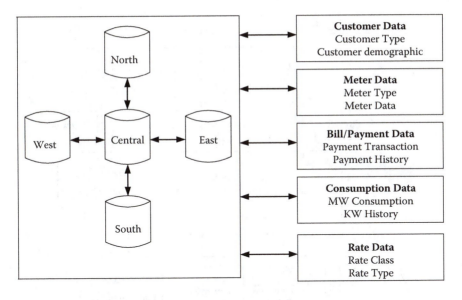

Figure 6.6 Data concepts in a CIS.

Electric utilities have tough competition among each other due to the deregulation in process and customers choosing to select the preferred supplier of electricity. Hence, it is important for the distribution utilities to enhance the CIS functionalities to attract and appeal to more customers and expand the business. The new functionalities may utilize advanced data warehousing and intelligent data mining techniques to build and establish customer profiles, consumption patterns, and predictions of customer preferences and behavior.

The power of data warehousing, as explained in Section 4.10 and shown in Figure 4.20, can be utilized to provide better services by adopting customer relations management (CRM), which will enable the utilities to build healthy relationships with customers, improve the level of satisfaction, and identify preferred customers. A typical CIS data warehouse with customer data information is shown in Figure 6.6.

6.3.4   GIS (geographical information system)

A geographic information system (GIS) captures, stores, manipulates, analyzes, manages, and presents all types of geographical data. GIS provides spatial data entry, retrieval, and visualization facilities.

GIS application to distribution automation is very important, as development of a geo-referenced network and customer database is necessary for proper integration of the distribution functions. The power distribution companies constantly engage in updating their consumer data and

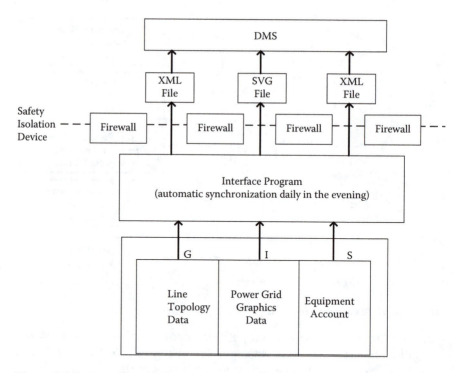

Figure 6.7 Data exchange between GIS and DMS.

the corresponding electrical network attributes. The mapping of the consumers' and network's assets will help the utility to define the network
connectivity with the physical location. The entire electrical network will
be laid over the satellite image or a vector-based map with the facility for
zooming, panning, scrolling, and resizing.

GIS is integrated with CIS, OMS, AMS, and billing and accounting
systems of the utility. It is also used for distribution planning studies, load
flow, and load management.

A typical GIS integration with DMS is shown in Figure 6.7 where
the line topology, graphics data, and equipment account information are
transferred to the DMS in different file formats.

6.3.5   AMS (asset management system)

Asset management is a business philosophy designed to align corporate
goals with asset-level spending decisions. It is one of the most important
activities of an electric utility for reducing the risk of eventual failures and
ensuring good performance of the assets. The assets of an electric utility
are the equipment installed in the field, the transformers being the most
expensive assets, insulators, bushings, conductors, and associated switch

gear. Hence, asset management may be defined as the process of maximizing equipment investment return by maximizing performance and minimizing cost over the entire equipment life cycle. AMSs may reduce expenditure, manage risks more efficiently, or drive corporate objectives throughout the organization. Recent developments show that asset management is an art of balancing cost, performance, and risk involved. Public image of the utility can be a deciding factor while framing the asset maintenance policy because social impacts in today's world are tremendous. The duration and frequency of supply failures and the resulting public image of the utility will depend on the type of maintenance policy adopted by the utility. Thus, there is a need to align the maintenance policy with the utility's corporate and strategic goals.

Figure 6.8 depicts the decision-making process according to an asset management point of view. It is possible to observe some aspects usually considered in this approach, such as equipment condition assessment, reliability management, and risk analysis, taking into account the data and information flow from the component level up to the corporate level.

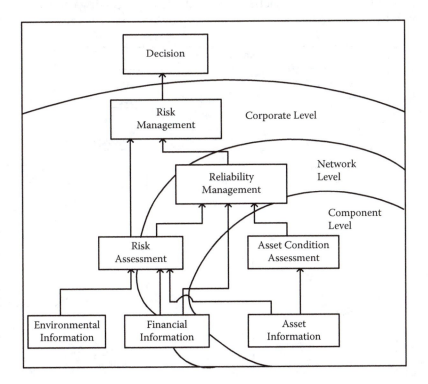

Figure 6.8 Decision-making process in an AM approach. (Source: Cigre WG 23/19-14, 2002.)

6.3.6   AMI (advanced metering infrastructure)

Automated meter reading (AMR) played a major role in simplifying and enhancing the meter reading challenges of electric utilities; the cost of meter reading was reduced and efficiency improved. Today's customer demands and aspirations require a two-way communication between the utility and customer and an advanced metering infrastructure (AMI) is can gather meter data and also use the energy meter for an array of functionalities for smart grid implementation. AMI also enables the utility to execute control functionalities and gives the customer a faculty to choose the electricity usage pattern and tariff structure and thus control electricity bills.

Figure 6.9 is a block diagram of a typical AMI system, which includes the collecting unit, communication network, and data processing center, generally referred to as the meter data management system (MDMS). The collecting unit collects data from the meter and other devices such as the switches and will distribute any control signal received from the control room to the customer meters. The AMI communication infrastructure will link the data collecting unit to the data processing center, which generally is located in the substation or control room. It may be noted

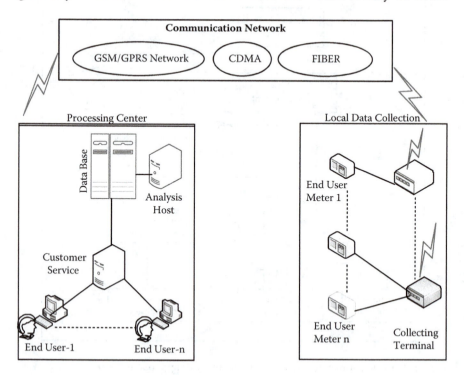

Figure 6.9 AMI infrastructure.

that the meter, data collection unit, and communication network will have bidirectional communication facility. The data processing center will be part of the CIS which disseminates data to the customers via a Web portal with log-in facilities. AMI will be discussed in detail in Chapter 7.

6.4 DMS framework: Integration with subsystems

It is evident that distribution management systems work seamlessly with other subsystems in the control center, and the data flow can be represented in Figure 6.4. The SCADA/DMS system is connected to an integration bus that distributes relevant SCADA/DMS data for use by the OMS, AMS, CIS, GIS, and AMI. Similarly, the DMS system acquires the required data from OMS, AMS, CIS, GIS, and AMI via the enterprise integration bus.

The optimization of the system operations is achieved in this way by coordination of the different systems. Figure 6.10 shows an overview of the integration of the components of a distribution automation system where the data from the substations are carried to the control center by the communication infrastructure. The control center has SCADA functionalities for online monitoring and control of the system, and accommodates the DMS, OMS, and MDMSs for AMI implementation. The consumers

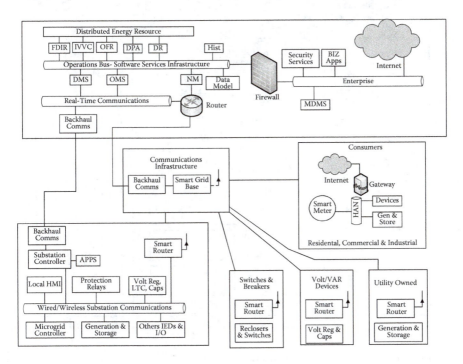

Figure 6.10 (See color insert.) DMS integration.

of electricity are also connected to the communication infrastructure to enable bidirectional communication. The smart switches, breakers, voltage regulators, capacitor banks, and any utility-owned generation and storage facilities are also hooked to the communication infrastructure for online monitoring and control as part of the distribution automation. The diagram also emphasizes the optimization of infrastructure, resources, and data inputs for the different functionalities in the control center, as discussed.

Figure 6.11 shows the framework of a DMS system which is adopted by industry, where the SCADA/DMS works in unison with data inputs from GIS, AMI, and OMS. The OMS gets inputs from CIS, AMI, and interactive voice response (IVR). The figure also shows the works management and movement of the mobile workforce, thus completing the cycle in case of an emergency. The SCADA/DMS consists of the SCADA platform, the network model that drives on-time applications (e.g., distribution load flow) and network analysis applications (e.g., integrated volt-var control). The application functions in the central block are explained in Section 6.5.

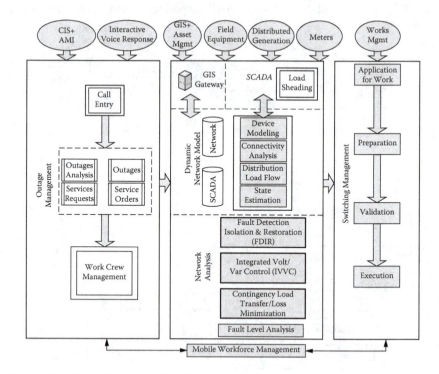

Figure 6.11 (See color insert.) DMS framework.

6.4.1   Common information model (CIM)

Distribution companies often use different information technology (IT) platforms from separate vendors to implement functions like OMS, AMI, and GIS. These systems generally use proprietary formats for data exchange, and ensuring the data exchange between different systems is a formidable task. Developing special adaptors for this purpose consumes a lot of money and manpower in a utility IT department. The common information model (CIM) is an attempt to establish a uniform language and domain model for energy management systems and related data structures. CIM builds an integrated platform to connect the different applications in a distribution utility and is defined in IEC 61970. The core package defining the CIM is IEC 61970-301. IEC 61970-501 and IEC 61970-452 define XML format for network model exchange. IEC 61968 defines the distribution-related data integration for DMS, OMS, AMI, GIS, CIS, and planning. These CIM models are depicted in Section 3.9.9 and Figure 3.16.

6.5   DMS application functions

The distribution management system runs a number of application functions to assess the distribution system status and also the performance so as to help the operator make intelligent and appropriate decisions. The application functions can be categorized into real-time application functions and analytical application functions. Real-time application functions performed help the operator to keep the system balanced and also deliver quality uninterrupted power to the consumers. The analytical application functions provide the inputs for optimizing the operation of the system by performing coordination functions and optimal feeder reconfiguration and capacitor placement algorithms. The DMS is also linked with other subsystems within the distribution utility such as outage management systems and customer information systems. The following sections will explain the application function in detail. Figure 6.12 a gives an overview of the application functions and demonstrates the interdependence of the functions and the advantages to the operator and utility at large. Important application functions will be explained in brief in the following sections.

6.6   Advanced real-time DMS applications

6.6.1   Topology processing (TP)

The distribution systems are vast networks running across the neighborhood, and the topology of the network often changes according to the

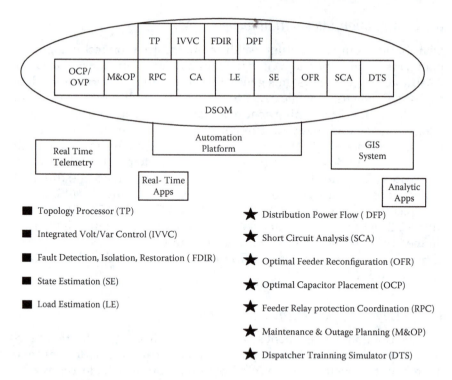

Figure 6.12 DMS application functions.

operating conditions of the system. The opening and closing of switches and faults on the system will alter the network, and building the grid connectivity is a challenge. The connectivity of the branches and node devices in the field, the service status of the branch devices, and the open or close status of the switching devices are reflected in the topology processor. The topology processor will indicate the grid connectivity for problem formulation for network analysis and optimization applications, dynamic network visualization by color changes, intelligent alarm processing, and network management. Different solution methods can be used for ensuring the topology update on the system such as wide-first search, deep-first search, or the bus-merging search.

6.6.2   Integrated volt-var control (IVVC)

Capacitor compensation is the best way to provide reactive power to the load. Capacitive compensation also improves the voltage profile of a feeder and at the same time reduces the feeder losses. The power factor correction is an added advantage of the capacitor placement. Capacitors are of two types: fixed on the feeder, and switched on and off as per the requirement. The objectives thus of integrated volt-var control are to minimize losses, maximize feeder power factor, minimize feeder power consumption, and maintain a flat voltage profile during all load conditions (high, medium, or low). The control variables are as follows:

1. The feeder capacitor bank on/off with remote control
2. Feeder voltage regulator tap position with remote control
3. Substation transformer tap position with remote control

This function utilizes single interval optimization, as per the interval chosen, and can compute the volt-var values in real time for the current load levels and also for specific future planning.

Different scenarios can be programmed like "minimize real power consumption" for peak loading of the feeder or "improve power factor" for a medium load scenario.

A typical feeder volt-var control scheme is shown in Figure 6.13.

6.6.3 Fault detection, isolation, and service restoration (FDIR)

Fault detection, isolation, and service restoration (FDIR) are very important application functions that help the utility to improve reliability of service. The main objectives of this function are fast service restoration, maximization of the number of customers restored, and minimization of the number of switching operations. These actions cause minimum impact to nonfaulted sections, and should also ensure easy return to the prefault status after the fault is repaired.

The control operations involved are the opening and closing of feeder tie switches and feeder line switches with remote control, as shown in Figure 6.14. This function is used in dealing with emergency conditions and will depend completely on the current circuit configuration and loading conditions. It is observed that in an automated system, the service restoration time reduces from 75 min to 1 to 5 min with FDIR implemented for customers on the healthy sections of the feeder (Figure 6.14).

The FDIR can be demonstrated by the example given in Figure 6.15, where there are five feeders supplying power to a distribution network. Feeder 4 is supplying maximum power as the lower part of the network is fully supplied by this feeder as can be seen with switches S1, S2, S3, and S4 normally open.

When a fault occurs on feeder 4 as shown, the following FDIR scheme can be implemented:

- A fault occurs at feeder 4.
- The CB 4 trips, disrupting supply to feeder 4 and associated lines which include the lower part of the network.

Feeder VVC Control

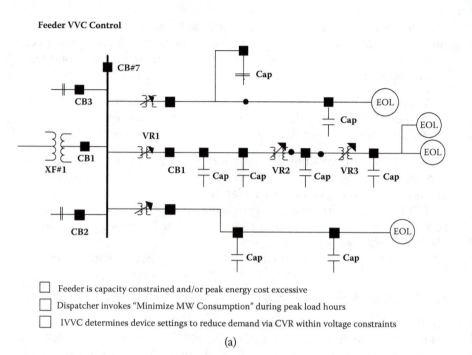

☐ Feeder is capacity constrained and/or peak energy cost excessive

☐ Dispatcher invokes "Minimize MW Consumption" during peak load hours

☐ IVVC determines device settings to reduce demand via CVR within voltage constraints

(a)

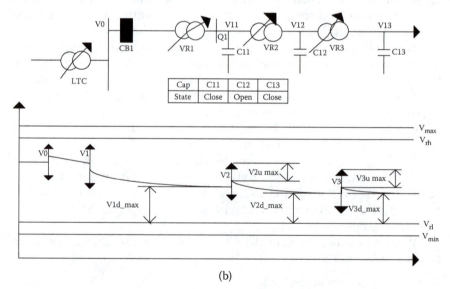

(b)

Figure 6.13 (a) Feeder voltage control. (b) Capacitive voltage regulator in IVVC.

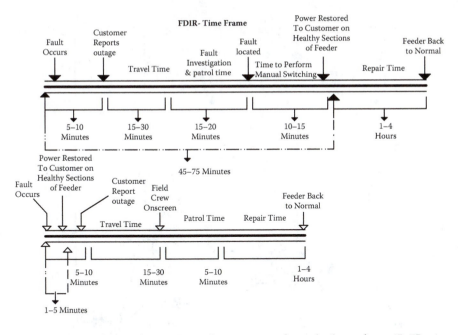

Figure 6.14 FDIR time frame, time of restoration of supply down from 45–75 min to 1–5 min.

- Fault is located by suitable algorithm.
- Fault is isolated by opening switches S5 and S6 by a command from the control room.
- CB 4 is closed, restoring supply to part of feeder 4 until S5.
- Switch S7 is opened to break the lower part of the network into two sections.
- Switch S4 is closed to restore supply to part of the interrupted portion from feeder 5.
- Switch S2 is closed, restoring supply to the rest of the network from feeder 2, thus completing service restoration.
- The above FDIR can be achieved in 1 to 5 min or less, and the consumers between switches S5 and S6 where the fault has occurred will get power after the crew fixes the permanent fault on the line.

FDIR is categorized into manual, semi-automatic, and fully automatic approaches.

In the case of the *manual approach*, there is no automatic control—the FDIR system delivers recommendations of switching schemes to the operator who executes the recommended actions. This approach is simpler than fully automatic, and it also acts as a confidence-building approach for the application logic and its recommendations. However, the flip side is that it

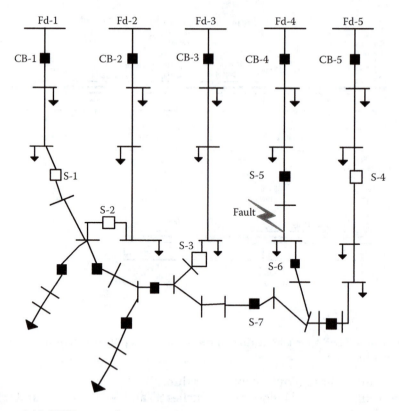

Figure 6.15 FDIR example.

takes a longer time to restore the service which is typically greater than 5 min depending on the two-way communication time and the time taken by the dispatcher to make an appropriate decision. It will also be difficult for the operator to manage multiple switches in case of an emergency with multiple disturbances.

The *semi-automatic approach* is a combination of automatic and supervised scenarios, where the system automatically isolates the fault and performs *upstream* restoration—that is, restores the supply from the substation to the node above the fault. The *downstream* restoration activities are performed manually by the dispatcher based on the recommendations by the system. The advantages of this approach are that it is simpler than fully automatic and is a natural progression from the manual approach. This is the scheme adopted by most utilities. In this case, the upstream customers get restored service in less than a minute. The disadvantage is that it takes longer to restore the downstream customers, and it will also be difficult for the operator to manage multiple switches in case of an emergency with multiple disturbances.

In the *fully automatic approach*, the fault isolation and restoration activities are fully automated without any human intervention. In this case it is possible to restore all service in less than 1 min and there will be less burden on the dispatcher to manage the switching activities. However, it is the most complex approach, and it is sometimes difficult for conventional dispatchers to accept the fact that the system restoration occurs without their intervention.

6.6.3.1 FDIR control strategies

The control of the FDIR scheme can be implemented in many ways, using stand-alone automatic switches like reclosers and sectionalizers with manual control by the dispatchers, or the control can be centralized from the SCADA/DMS system in the central control room. Yet another method of control is substation centered where the logic in a substation controller controls the associated feeders. The peer-to-peer approach utilizes the communication facility between switches where the group of switches communicates with each other and determines the switching actions.

6.6.3.2 Reliability indices

Reliability indices for power distribution utilities are defined in IEEE standard P1366, "Guide for Electric Distribution Reliability Indices." The most common distribution indices include the System Average Interruption Duration Index (SAIDI), Customer Average Interruption Duration Index (CAIDI), System Average Interruption Frequency Index (SAIFI), Momentary Average Interruption Frequency Index (MAIFI), Customer Average Interruption Frequency Index (CAIFI), Customers Interrupted per Interruption Index (CIII), and the Average Service Availability Index (ASAI). Some of the indices are defined below, and these indices serve as indicators of the system reliability.

SAIFI = Total number of customer interruptions/total number of customers served.

SAIDI = Total duration of customer interruptions/total number of customers served.

CAIDI = Total duration of customer interruptions/total number of customer interruptions.

MAIFI = Total number of momentary customer interruptions/total number of customers served.

CTAIDI = Total duration of customer interruptions/total number of customers interrupted.

CAIFI = Total number of customer interruptions/total number of customers interrupted.

ASAI = Customer hours service availability/customer hours service demand.

ASIFI = Connected kVA interrupted/total connected kVA served.

ASIDI = Connected kVA duration interrupted/total connected kVA served.

FDIR schemes are extensively used, and the reliability indices show tremendous improvement using these schemes.

6.6.4   Distribution load flow

Distribution systems are poorly conditioned power systems that are mostly radial with only a few loops in some cases. Hence, the admittance matrix and Jacobian matrix formed will be increasingly sparse. The distribution lines have high resistance to reactance (R/X) ratios due to many reasons. The distribution system also has to be modeled as a three-phase system as the grid is unbalanced due to the presence of single-phase, nongrounded, and grounded supply combined with unbalanced loads and distributed generating sources. The network model will be large due to a large number of nodes in the distribution system. Normal load flow solutions used in transmission networks utilize a fuller Jacobian matrix and have a low R/X ratio, and a single-phase model of the system could be used because at the transmission level, the system is balanced. The Newton-Raphson, Gauss-Siedel, or fast decoupled load flow techniques are used to solve the transmission system load flow; however, these techniques fail to converge in the case of a distribution system due to the reasons given earlier. The distribution systems use different load flow solutions, as they are poorly designed with high R/X ratios. Some commonly used techniques include backward/forward sweeping algorithms using Kirchhoff current law (KCL) and the Kirchhoff voltage law (KVL), load flow based on sensitivity matrix for mismatch calculation, and the bus impedance method.

The distribution load flow computes the node voltages, branch current flows, and complex power flow in the system. There could be changes in the possible grid condition due to the variation in power flow from specific generation sources, load changes, capacitor bank insertions, and load tap changer operations, and the load flow program needs to take care of these factors. In real time the load flow solutions are sought by a dispatcher to verify the grid operating conditions with the known node injection real and reactive power telemetered in real time.

6.6.5   Distribution state estimation (SE) and load estimation

State estimation is the process of assigning a value to an unknown system variable based on measurements from that system according to some criteria. Usually the measurements from the power system may have errors; hence, the state estimator uses statistical methods and optimization techniques to arrive at the best estimate of the system states from available measurements and system topology. State estimation is well established for the transmission system and is widely used in all transmission SCADA control centers.

Distribution state estimation (DSE) computes the minimum set of variables that uniquely determine the grid operating conditions, node voltage magnitudes, and phase angles, along with the node and branch real and reactive power flows. This essentially leads to the computation or rather estimation of the loads at different points of the distribution system. DSE aims to compute an accurate estimate of system loads, which are used as inputs to modern DMS applications such as optimal volt-var control and optimal feeder reconfiguration.

State estimations in transmission systems are well established, whereas DSE has several challenges to overcome such as

1. A limited number of real-time measurements, many of which are current magnitudes
2. An unsymmetrical network design coupled with an unbalanced load operation
3. A large resistance to reactance ratio particularly for underground cables
4. The large sizes of distribution networks.

Therefore, the methodologies of transmission system state estimation are not applied as such to distribution systems. When the distribution systems are predominantly radial, state estimation is practically equivalent to load estimation. This has motivated approaches that offer a compromise between conventional load allocation and a fully sophisticated AC state estimator. Another class of distribution system state estimators is based on the theory of transmission system state estimation but uses load data as pseudo-measurements. Several practical distribution system state estimators have been built around load flow algorithms, where a statistical estimation method like weighted least square (WLS) is used, generally on a reduced network to get estimates of the states, followed by the distribution load flow of the full system.

State estimation holds key to many decisions as it provides many insights into the state of the distribution system. AMI gives a boost to DSE as it can provide a huge number of near real-time data at every service connection point, hence enhancing the accuracy and quality of the state estimation.

6.7   Advanced analytical DMS applications

6.7.1   Optimal feeder reconfiguration

Distribution systems operations must be adequately planned to permit efficient and reliable operation. With the exception of some urban systems, the majority of distribution systems operate with a radial topology for various technical reasons, the two most important being to facilitate the coordination and protection and to reduce the short-circuit current of distribution systems. Thus, the radiality constraint is a factor in most expansion and operation planning problems. The most widely known problem is the distribution system feeder reconfiguration, and the main objective is to find a radial system generating minimum losses.

The reconfiguration is done by changing the status of normally closed sectionalizing switches that exist between each section of an individual feeder and normally open tie switches that connect sections of different feeders. Network reconfiguration is important for both distribution automation and network planning. As the operating conditions vary, network reconfiguration can be used to minimize power losses provided that technical operational limits are not violated and protective devices remain properly coordinated.

From an optimization perspective, network reconfiguration is a mixed binary nonlinear optimization problem where binary variables represent the switch statuses and continuous variables model the power network. Although the globally optimal configuration can in theory be obtained by enumerating all feasible radial configurations and choosing the one with the least power loss, the process becomes impractical even for moderate size networks.

Figure 6.16 shows the IEEE 33 bus distribution network with the dotted lines representing the branches that can be brought into the network, as per the system requirements for optimal operation, by opening and closing the sectionalizing and tie switches.

6.7.2   Optimal capacitor placement

Installation of capacitors in primary and secondary networks of distribution systems is one of the efficient methods for energy and peak load loss reduction. Also, the voltage profile in the feeder is improved and static voltage stability is enhanced. The main challenge is the determination of optimal locations and sizes of fixed and switchable capacitors with respect to network configuration, distribution of load in the feeder, time variation of load, and uncertainty in load forecasting or the load allocation process.

There many techniques used by researchers for optimal capacitor placement; however, the rule of thumb generally used by industry is to

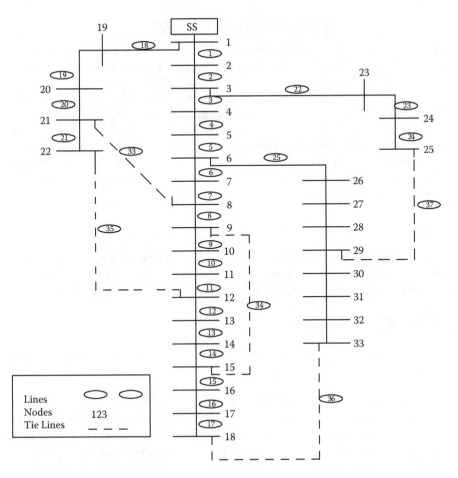

Figure 6.16 IEEE 33 bus distribution system.

use two-thirds of the feeder reactive requirement at two-thirds of the feeder length.

Network reconfiguration and capacitor placement have been widely employed to reduce power losses and maintain voltage profiles within permissible limits in distribution systems.

6.7.3   Other applications

Many other analytical applications are performed by the DMS such as short-circuit studies followed by feeder relay protection coordination. Power quality analysis can be easily performed by analyzing the data acquired from telemetry, operational, and nonoperational data in the data warehouse, as discussed elsewhere.

6.8   DMS coordination with other systems

Distribution management systems have to be integrated with many other subsystems in the substation and/or master control room of a distribution utility. The data available with the SCADA/DMS systems are useful to many subsystems, and vice versa, and hence this coordination is the key to a successful distribution automation implementation.

Historically, the DMS data sources were the SCADA data and the customer trouble calls and maintenance crew reports. However, there are many automated application subsystems in the distribution control center which can help the DMS perform better, such as the outage management systems (OMS), GIS, and AMI.

6.8.1   Integration with outage management systems (OMS)

Table 6.1 indicates the coordination between OMS and DMS, and it is clear that both the subsystems benefit from the coordination, and it is essential to interlink the information so that optimum benefits are derived.

6.8.2   Integration with AMI

AMI can provide a large number of inputs to the DMS, of which a few are listed below.

6.8.2.1   Consumer energy consumption data

The AMI record the energy consumption of every customer at an interval of 5 to 15 min depending upon the time interval set for energy billing. It also provides the maximum and minimum kW power levels in each interval and the energy generation by each customer (if any) in the interval. The AMI also gives information about the daily consumption profiles of customers.

Table 6.1 OMS Coordination with DMS

OMS		DMS
Update grid topology from operator and automatic controls	⇐	Switch operations from operator and automatic controls
Estimated fuse or manual switch statuses based on trouble calls	⇒	Update the determined fuse and switch statuses
Predict customer outages and impacts based on the DMS plan	⇐	FDIR and other planned switch operation schedule
Make switch orders and estimate repair time for DMS operation	⇒	Execute the automatic switch orders from OMS

Table 6.2 AMS Integration with DMS

AMI	DMS
Aggregated energy data for each interval in each user transformer	Form load profiles for individual loads (user transformer)
Aggregated reactive energy data for each interval in each user transformer	Perform state estimation or load estimation
Aggregated voltage profile data for each interval in each user transformer	Analyze load voltage profiles for volt-var control and voltage optimization
Consumer energization status in each interval in each user transformer	Outage detection from the user energization statuses

6.8.2.2   Reactive energy consumption

The reactive power consumption during each interval and the maximum and minimum voltage levels at each interval are recorded by the AMI.

6.8.2.3   Voltage profile data and energization status data

The AMI will provide the voltage profile of the customer and feeders for each sampling interval and at the same time provide the energization status at the level where it is required, whether at the transformer level or bus bar level or at the customer level.

Table 6.2 elaborates the integration of AMI with DMS for better management of the distribution systems.

6.9   Customer automation functions

Customer automation focuses on smart meters installed at the customer premises and the advanced metering infrastructure (AMI) that operates as two-way communications between the customers and the electric utility. The customers are encouraged by the utility to participate in the optimal use of energy, delaying additional investment for the utility and reducing the electricity bills of the customer, where load management by the utility has now shifted to a demand response mechanism. Customers are installing home energy management systems to guide them in utilizing energy in the most efficient way. In addition, large utilities are establishing elaborate customer information systems and trouble call management systems, with interactive voice recording facilities signaling the roles of the customers in the distribution management. Smart grid implementation is a way of encouraging customer participation in the distribution management and will be discussed further in Chapter 7.

6.10 Social media usage for improved reliability and customer satisfaction [17]

Reliability will be the top priority for grid operators no matter whom or where they serve. From a tree branch that takes out a single home to a complex, multifactor event that cuts power to 700 million people (as a two-day event did in India), rapid outage detection and speedy restoration play critical roles.

Regulators assess utilities' performances partly on reliability indices sometimes referred to as "SAIDI (System Average Interruption Duration Index) and its sisters": SAIFI (System Average Interruption Frequency Index), MAIFI (Momentary Average Interruption Frequency Index), and CAIDI (Customer Average Interruption Duration Index). It is challenging to keep these numbers within acceptable limits; they are based on IEEE Standard 1366.

As intelligence is extended down into the distribution system, much data can be generated if used properly by an integrated system of outage detection and power restoration and can help utilities improve critical indices. This can be done using available technologies. In addition, vendors are developing software applications that can monitor social media to provide speedy, granular data to assist these efforts. The attractive secondary benefits are customer engagement and satisfaction.

These developments and their application can provide utilities with a clear path through the smart grid woods and maintain their focus on the most important factor that guides their work: power grid reliability.

In describing how to improve outage detection and power restoration, there is an integrated approach to distribution automation that takes advantage of the strong business case involving fault detection, isolation, and restoration (FDIR), and integrated volt-VAR control (IVVR). For utilities that have installed advanced metering infrastructure (AMI) and interval smart meters, this is a logical step; however, AMI is not necessary to reap the benefits of the approach outlined here.

As discussed earlier, automated substations that can trigger switches to isolate faults without operator intervention still leave utilities to rely on customers to call about outages. That is why advocating a more integrated approach to full distribution system automation will improve multiple benefits for the business case and utility performance on outage-related indices. Distribution automation's business case improves markedly if the driving focus is on speedy outage detection and power restoration.

Smart meters record electricity usage for billing, measure end-of-line voltage, and in the case of an outage, emit a last gasp as they lose power. Capacitors are designed to hold enough charge to get that crucial message out. That is almost instant, more precise, and quicker than waiting for customers to call—sometimes the difference between an outage falling under MAIFI, a momentary outage, and SAIFI, a sustained outage.

For utilities without interval meters and AMI, voltage-sensing meters can be placed strategically at the ends of feeders to ensure compliance with national standards for delivering a standard voltage range. For example, in the United States, the American National Standards Institute (ANSI) redefines the voltage at the delivery point to the customer to be 120 V ± 5% (114 to 126 V). Those meters can play a role in outage detection, though a less granular one than full AMI metering.

A traditional customer phone call could be linked to a physical address by tapping into the customer information system. With the widespread use of social media, the customers' tweets can be linked to an address to obtain similar information. This can be done in various ways. The utility could incentivize customers to link their Twitter tags to their account information or to turn on their mobile devices' geo-tagging functions, which provide latitude and longitude that indicate their locations when they tweet instead of call. A cluster of tweets tied to addresses—the greater the sample, the greater the accuracy—can be subjected to automated analysis to provide the precise location and extent of an outage. *Grid IQ Insights* is a new software platform with geospatial coordinates for automated systems that can connect tweets with an outage management system.

Getting customers to work directly with their utilities takes time and effort. Customers are more likely to tweet each other to complain of outages. The new software application uses text mining to understand whether a flurry of tweets that mention "outage" and "power" really refer to a "power outage," and if so, can mash up that data with end-of-line sensor data to identify the location and extent of an outage rapidly. Because each household has multiple members and an electricity account holder might not be the one tweeting, this method provides an attractive alternative for utilities without AMI. The analytics software that determines the location of outages in both instances resides in the utility's outage management system (OMS) platform but remains a separate function.

6.10.1 Replacing truck rolls

Crowd sourcing the cause of an outage by leveraging customers' use of still and video camera capabilities on ubiquitous mobile devices is another social media benefit to utilities. Incentivizing customers to post their pictures and videos of downed lines to a utility's Facebook page while emphasizing traditional cautions about approaching high-voltage situations could provide images to be analyzed by automated image-assessment software to better inform field crews prior to truck rolls.

Even where AMI is installed, there is only 1 meter per household. Data show that each household has multiple social media accounts, potentially multiplying the number of resulting data points. The current 200 million Facebook accounts in the United States represent about one account per household, and that number is growing. Using social media, as parents know, often is preferred to making phone calls. An app could reduce outage notification to one click. As generations shift, these trends will become more pronounced.

We see in these trends immediate operational benefits tied to outage detection and power restoration plus the seeds of significant customer interaction and the larger field of general consumer engagement. Educating consumers about the economic and societal benefits of a smarter grid will be the first step in creating a smart grid social network. That network will function much like the smart grid, with open, collaborative, two-way information flow between consumers, the ultimate deciders of smart grid, and utilities, the ultimate providers of smart grid. That interaction will lead to customers' education in and acceptance of a new utility-customer paradigm in which customers understand and willingly participate in critical utility programs such as demand response. That participation will signal a human smart grid, and it should reflect customers' understanding of electricity's role in the macro economy, as well as their own lives and prosperity.

Once consumers in general grasp that energy efficiency, demand response and active energy management have positive implications for their pocketbooks, economic security, energy independence, and the environment, they will become the utility's partners in achieving those goals. As utilities seek to defer new capital investment where possible, wring efficiencies from the grid, and move from fossil fuels to more renewable and sustainable sources, they will find that an educated, cooperative customer is one of their most valuable resources. Utilities will learn to be customer savvy, responsive, and conscious of opportunities to add value to their commodity and its delivery.

6.10.2   Tying it all together

Now we have outage data, either through AMI or end-of-feeder sensors, plus social media. How those data are routed and analyzed is critical to speedier power restoration. A well-designed communication network will enable smart meters' last gasps to trigger near-instant automated switching. Without AMI, but by combining end-of-line sensor data and social media data, operators play an active role. Outage detection and power restoration will be speedier than relying on customer phone calls, especially if the applications run closed loop and remove the operator from the decision-making process.

Automation in this case has three components: a control center master, field equipment, and a communication network. All three factors impact performance and cost. The advent of public networks, mostly wireless, has provided the tipping point for the distribution automation business case. It is finally cost effective to provide a reliable data network across a large geographic area, whether that network is public or private. Communication network performance is measured by three variables: response time, bandwidth, and latency. Utilities' performance requirements allow different response times for different applications, such as switches needing 2 s, analog information needing 15 to 30 s, and a capacitor bank that can be triggered in 30 s. Bandwidth is measured in bits per second. How much data and how quickly can the sequence of events be reported? Latency refers to how much delay is acceptable in transmitting or receiving signals.

The network design must balance overhead on the system with the speed needed for various signals. For instance, cybersecurity and cryptographic requirements will introduce latency into the signal path; a 200 to 300 ms latency can become a 600 to 800 ms latency when cryptography is applied. Industry standard communication protocols (e.g., DNP3) introduce overhead, as well, and the communications system bandwidth might need upgrading to maintain the same update rate at the control center master.

Most utilities will employ a hybrid communication network. At larger substations, utilities might use fiber-optic cable or licensed wireless frequencies. Smaller, peripheral substations require only unlicensed spectra for wireless radio. Downstream of the substation, the solution likely is a wireless private network, considering response time, bandwidth, and latency requirements, as well as cost.

6.10.3   Routing signals

Another challenge is to integrate outage and verification messages from the meters or end-of-line sensors through the substation to the control center and its OMS. AMI systems typically are designed to support only meters' interval data, which travel from the meter to the head-end system to a meter data management system (MDMS) for storage and analysis. The AMI may not be designed for voltage data or the meters' last gasps, which need to be routed around the AMI's path and directed into a distribution management system (DMS) or an OMS. The DMS will use voltage data to populate a network model, and an OMS is the proper destination for last-gasp signals.

These separate paths reflect the distinction between operational data and nonoperational data, and proper data routing around the AMI system is a nascent functionality that utilities must demand from vendors.

Utilities should focus on operational data in this context, but vendors must enable utilities to extract more value from the nonoperational data coming out of IEDs, which will provide value to asset management, maintenance, and power-quality efforts.

6.10.4 DMS in outage management

Consider the DMS's role in an integrated system. DMS relies on a network model generated from geographic information system (GIS) data and is populated by substation and feeder intelligent electronic devices and voltage data from end-of-line sensors. A network model manager interfaces with a GIS so it knows what data to pull from the GIS to build a three-phase, unbalanced DMS network model.

A utility needs four applications on a DMS: the aforementioned FDIR and IVVR, optimal feeder reconfiguration (OFR), and distribution power flow (DPF). Protective relays detect a fault, its location, and its type. Then the FDIR isolates the faulted segment of the feeder and restores power to customers on healthy segments of the feeder using the OFR. An OFR can look ahead to account for switching schedules for routine maintenance to optimize its role.

All this should happen in less than 5 min, keeping an event within the MAIFI index for most customers and not impacting SAIDI, SAIFI, and CAIDI.

Although IVVC is only incidental to outage management, it plays a big role in DMS, optimizing voltage and reactive power for energy efficiency.

The DPF is an online tool that allows the operator to simulate the results of switching strategies and thus contributes to energy efficiency by controlling losses and loading on feeder lines. These two related functionalities make an outsized contribution to the business case for integrated distribution system automation. The suite of functionalities available through FDIR and IVVR comprises the best business cases for adding intelligence to distribution systems.

Advancements in data visualization tools such as dashboards make all the interactions described here graphically clear to the grid operator in the control center, and in circumstances requiring operator action, give data in the form of actionable intelligence.

With all the aforementioned elements in place plus that crucial two-way communication with the customer, utilities can improve their reliability indices and begin to engage customers in a virtual cycle that further contributes to speedier outage detection, power restoration, and customer satisfaction.

Utilities must weigh the larger value of increased reliability and its impact on customers, regulators, and other stakeholders.

6.11   Future trends in DA and DMS

Distribution automation revolves around the automation of substations, feeders, and customers in a distribution system. With the advent of cheaper automation products and better two-way communication options, automation of distribution systems is progressing well all over the world. This opens a new vista of applications to optimize the assets, demand, transmission, and distribution systems, workforce, and engineering in an electric utility. Demand-side management and demand response, integration of renewable energy sources, energy storage, smart homes, and energy management systems are areas of expansion, and at the same time are areas of concern. Electric vehicles and their charging stations are going to pose problems and challenges for distribution systems. Microgrids with energy generation capability are the next big investment by utilities. These future trends in distribution SCADA are discussed in detail in Chapter 7.

6.12   Case studies in DA and DMS

Utilities have realized the benefits of implementing DA and DMS functionalities, and the following case studies present different perspectives.

On the Road to Intelligent Distribution [34]. The remote operation and access to the distribution system (ROADS) project aims at improving the visibility and control of distribution systems for Duke Energy. Three distribution substations were automated with continuous remote monitoring of power apparatus, with modern IVVC (integrated volt-var control) and transformer load tap changer control.

Reduction in Aggregate Technical and Commercial (AT&C) Losses in Tata Power-DDL Utility [35]. Tata Power DDL, Delhi, India, has implemented SCADA/EMS with grid substation automation system (GSAS) DMS and OMS successfully and a global standard call center service. By implementing integrated automatic demand response (ADR) and automatic metering infrastructure (AMI) projects, the company has been able to achieve an aggregate technical and commercial (AT&C) loss reduction of 12.5% against a target of 10.5% and a reliability of 99.5% at a peak of 1508 MW.

6.13   Summary

This chapter focuses on the distribution automation functions related to customer and feeder automation. The subsystem in the distribution control center such as the distribution management system (DMS), outage management system (OMS), customer information system (CIS), geographical information system (GIS), asset management system (AMS), and advanced metering infrastructure (AMI) are explained and their interdependency established. The chapter also discussed the DA application functions such as the topology processor, IVR, FDIR, and so forth, and the advanced analytical functions like optimal feeder reconfiguration, optimal capacitor placement, and other applications. The chapter also discusses social media usage for improved reliability and customer satisfaction.

Bibliography

1. H. M. Gill, Smart grid distribution automation for public power, *IEEE PES Transmission and Distribution Conference and Exposition*, pp. 1–4, April 2010.
2. Brad Tips and Jeff Taft, Cisco smart grid: Substation automation solution for utility operations, white paper, pp. 1–7, 2010.
3. John McDonald, Substation automation and enterprises data management to support smart grid, *G.E. Digital Energy, Georgia Tech Clean Energy Speaker Series program*, March 2011.
4. J. Sandeep Soni and Smita Pareek, Role of communication schemes for power system operation and controls, *International Journal of Electronics and Communication Engineering and Technology*, pp. 163–172, November 2013.
5. V. C. Gungor and F. C. Lambert, A survey on communication networks for electric system automation, *Elsevier Computer Networks Journal*, pp. 877–897, 2006, vol. 50, DOI. 10-1016. Connect 2006. 01.005.
6. Yves Chollot, Jean-Marc Biasse, and Alain Malot, Improving MV network efficiency with feeder automation, *CIRED—21st International Conference on Electricity Distribution, Frankfurt*, pp. 1–8, June 6–9, 2011.
7. Sandeep Pathak, Decentralized self healing solution for distribution grid automation: A must need of Indian distribution utilities, IEEE Innovative Smart Grid Technologies–Asia (ISGT Asia), pp. 1–6, November 10–13, 2013.
8. Joshua Z. Rokach, Smart houses in a world of smart grids, *The Electricity Journal*, vol. 25, no. 3, pp. 94–97, April 2012.
9. Zhuang Zhao, Won Cheol Lee, Yoan Shin, and Kyung-Bin Song, An optimal power scheduling method for demand response in home energy management system, *IEEE Transactions on Smart Grid*, vol. 4, no. 3, pp. 1391–1400, September 2013.
10. Eun-Kyu Lee, Rajit Gadh, and Mario Gerla, Energy service interface: Accessing to customer energy resources for smart grid interoperation, *IEEE Journal on Selected Areas in Communications*, vol. 31, no. 7, pp. 1195–1204, July 2013.
11. Farrokh Rahimi and Ali Ipakchi, Demand response as a market resource under the smart grid paradigm, *IEEE Transactions on Smart Grid*, vol. 1, no. 1, pp. 82–88, June 2010.
12. Robert J. Procter, Integrating time differentiated rates, demand response, and smart grid to manage power system costs, *The Electricity Journal, Elsevier*, vol. 26, no. 3, pp. 50–60, April 2013.
13. Shengnan Shao, Tianshu Zhang, Manisa Pipattanasomporn, and Saifur Rahman, Impact of TOU rates on distribution load shapes in a smart grid with PHEV penetration, Transmission and Distribution Conference and Exposition, 2010 IEEE PES, pp. 1–6, April 19–22, 2010.
14. Ali Ipakchi and Farrokh Albuyeh, Grid of the future, *IEEE Power and Energy Magazine*, March/April 2009, pp. 52–62.

15. Robert Smith, KeMeng, Zhaoyang Dong, and Robert Simpson, Demand response: A strategy to address residential air-conditioning peak load in Australia, *Springer Journal on Modern Power System Clean Energy*, vol. 1, no. 3, pp. 223–230, 2013.
16. Shengnan Shao, Manisa Pipattanasomporn, and Saifur Rahman, Grid integration of electric vehicles and demand response with customer choice, *IEEE Transactions on Smart Grid*, vol. 3, no. 1, pp. 543–550, March 2012.
17. John McDonald, Integrated system, social media improve grid reliability, customer satisfaction, *IEEE Smart Grid: Electric Light and Power*, December 2012.
18. D. Kreiss, Non-operational data: The untapped value of substation automation, *Utility Automation*, September/October 2003. Available at http://uaelp. pennnet.com/articles/articledisplay.cfm?article_id=192304.
19. J. Jung, C. Liu, and M. Gallanti, Automated fault analysis using intelligent techniques and synchronised sampling, TP 141-0, *IEEE PES Tutorial on Artificial Intelligence Applications in Fault Analysis*, July 2000.
20. A. H. Nizar, Zhao Yang Dong, and Ariel Liebman, Customer information systems for deregulated ASEAN countries, *The 7th International Power Engineering Conference*, IPEC, pp. 1–6, November 29–December 2, 2005.
21. T. H. Chen and Jeng-Tyan Cherng, Design of a TLM application program based on an AM/FM/GIS system, *IEEE Transactions on Power Systems*, vol. 13, no. 3, pp. 904–909, August 1998.
22. R. E. Brown and J. H. Spare, Asset management, risk and distribution system planning, IEEE PES Power Systems Conference and Exposition, vol. 3, pp. 1681–1686, October 10–13, 2004.
23. J. C. M. Lucio, J. L. T. Nunes, and R. C. G. Teive, Asset management into practice: A case study of a Brazilian electrical energy utility, 15th International Conference on Intelligent System Applications to Power Systems, ISAP '09, pp. 1–6, November 8–12, 2009.
24. M. Shahidehpour and R. Ferrero, Time management for assets—Chronological strategies for power system asset management, *IEEE Power and Energy Magazine*, pp. 32–38, May–June 2005.
25. I. Dzafic, M. Gilles, R. A. Jabr, B. C. Pal, and S. Henselmeyer, Real time estimation of loads in radial and unsymmetrical three-phase distribution networks, *IEEE Transactions on Power Systems*, vol. 28, no. 4, pp. 4839–4848, November 2013.
26. William Peterson, X. Feng, Z. Wang, S. Mohagheghi, and E. Kielozevski, Closing the loop, *ABB Review*, pp. 38–43, 2009.
27. Mathias Uslar, Tanja Schmedes, Abdreas Lucks, Till Luhmann, Ludger Winkels, and Hans-Jurgen Appelrath, Interaction of EMS related systems by using the CIM standard. DOI 10.1.1.73.5125.
28. Smart Grid, Smart City Program, Monitoring and Measurement Report IV, Grid Applications Stream: Active Volt-VAR Control, July 1–December 31, 2012.
29. Mini S. Thomas, Rakesh Ranjan, and Roma Raina, Fuzzy modeled load flow solution for unbalanced radial power distribution system, *Proceedings of the IASTED International Conference, Power and Energy Systems (EuroPES)*, Crete, Greece, June 22–24, 2011.

30. Mini S. Thomas and Parveen Poon Terang, Islanding detection using decision tree approach, *Proceedings of the IEEE International Conference: PEDES*, New Delhi, India, 2010.
31. John-Paul H. Knauss, Cheri Warren, and Dave Kearns, An innovative approach to smart automation testing at national grid, Transmission and Distribution Conference and Exposition (T&D), 2012 IEEE PES, pp. 1–8, May 7–10, 2012.
32. By Liang Che, Mohammad Khodayar, and Mohammad Shahidehpour, Only connect, *IEEE Power and Energy Magazine*, pp. 70–81, January/February 2014.
33. Ljubomir A. Kojovic, Craig A. Colopy, and Daniel Arden, Volt/VAR control for smart grid solutions, *CIRED 21st International Conference on Electricity Distribution*, Frankfurt, June 6–9, 2011.
34. R. Dougherly, S. Erwin, R. Uluski, and J. D. Mc Donald, On the road to intelligent distribution, *T&D World*, September 2006.
35. Annual report: Tata Power-DDL, 2013–2014, http://www.tatapower.com/investor-relations/pdf/TPDDL-2013-14.pdf.

chapter seven

Smart grid concepts

7.1 Introduction

High-quality electricity is a necessity in the modern world, due to the innumerable applications demanding quality power, such as electronic manufacturing, microprocessors, and many sensitive devices being used by common man. Hence, it is imperative of the electric utilities world over to supply affordable, reliable, and quality electric power to all [1].

In a traditional power system, centralized generating stations generate bulk power which is transmitted to the consumers through a one-way transmission and distribution system called the grid. Modernization of the grid is a priority for all the utilities, and governments are devising policies and practices driving the TRANSCO and DISCOs to modernization of the grid.

The motivation for modernization are manyfold with specific goals in mind [1,2]:

1. To make the production and delivery of electricity more cost-effective and efficient
2. To provide consumers with electronically available information and automated tools to help them make more informed decisions about their energy consumption and control their costs
3. To help reduce production of greenhouse gas emissions in generating electricity by permitting greater use of renewable sources
4. To improve the reliability of service
5. To prepare the grid to support a growing fleet of electric vehicles in order to reduce dependence on oil
6. To facilitate the integration of distributed resources into the grid and prepare the grid for the challenges involved
7. To delay investment intended to add capacity to generation, transmission, and distribution networks

Smart grid is the solution to the above concerns and will be discussed in detail in the following sections.

7.2   Smart grid definition and development

A smart grid delivers electricity from suppliers to consumers using digital technology to save energy, reduce costs, and increase reliability. It connects everyone to abundant, affordable, clean, efficient, and reliable electric power anytime, anywhere, providing a way of addressing energy independence and global warming issues.

Smart grid is a concept and may look different for different stakeholders. However, the envisioned smart grid concept will

- Motivate and include customers
- Resist attack
- Provide power quality for the 21st century
- Accommodate all storage and generation options
- Enable markets
- Optimize assets and operate efficiently
- Be self-healing

Smart grid is definitely the integration of the available electrical infrastructure with enhanced information capabilities, and it incorporates automation and information technology with the existing electrical network, so that the grid can operate in a smarter way. Smart grid implementations will provide comprehensive solutions that will improve power reliability, operational performance, and productivity for utilities. By making the grid smarter, energy use is managed efficiently, and customers will be able to save money without compromising on lifestyle. Optimal integration of renewables into the grid is a major benefit of smart grid implementation, and there will be substantial penetration of renewables in a smart grid scenario. Smart grid will provide meaningful, measurable, and sustainable benefits to all stakeholders by increasing energy efficiency and reducing carbon emissions.

7.3   Old grid versus new grid

The comparison of the traditional power grid and the smart grid (Table 7.1) will produce a host of advantages that the utilities, government, and customers will gain from the deployment of smart grid technologies. Figure 7.1 gives an overview of the old grid and migration to a smarter grid.

The conceptual diagram of the smart grid in Figure 7.2 shows the seven different domain entities and the mechanisms for successful operation of the new grid are clearly visible. The traditional bulk generation is transmitted and then distributed to the consumers; at the same time, the integration of distributed generation is done at the bulk as well as distribution and customer levels. The bidirectional information flow from

Table 7.1 Comparison of Traditional Grid and Smart Grid

Traditional grid	Smart grid
Customer calls when the power goes out	Utility knows when power is out and restores it automatically
Utility meets peak demand	Utility suppresses peak demand, thus lowering the cost
Difficult to manage high wind and solar penetration	Utility can manage distributed energy resources safely
10%+ power loss in T&D	Utility reduces power loss by 2+%, and reduce emissions and electricity bills of customers
All centralized control	Local control provides distributed generation

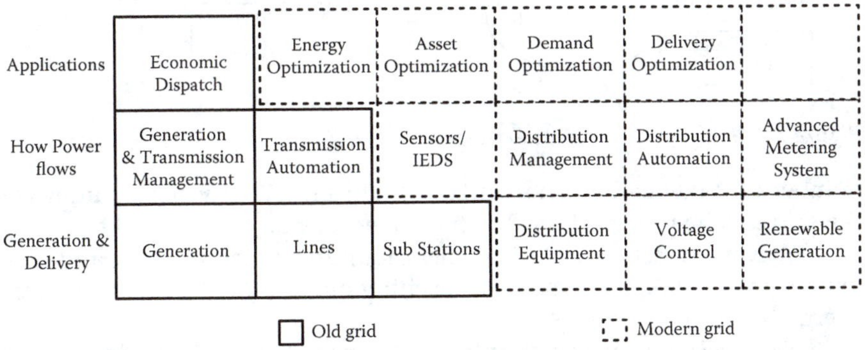

Applications	Economic Dispatch	Energy Optimization	Asset Optimization	Demand Optimization	Delivery Optimization	
How Power flows	Generation & Transmission Management	Transmission Automation	Sensors/ IEDS	Distribution Management	Distribution Automation	Advanced Metering System
Generation & Delivery	Generation	Lines	Sub Stations	Distribution Equipment	Voltage Control	Renewable Generation

☐ Old grid   ⌐ ¬ Modern grid

Figure 7.1 Migration of the old grid to a smart grid.

the customer to the utility by means of smart meters which enable real-time pricing is an added feature. The distributed storage at the utility and customer level coupled with electric vehicles and smart appliances at the customer premises loaded with networked sensors and home energy management systems adds value to the smart grid migration from the traditional grid. This has prompted the emergence of energy service providers and flexible energy markets with smart transmission phasor measurement units and wide-area monitoring and controls for a stable grid and for faster mitigation of any disturbance.

7.4   Stakeholders in smart grid development [3]

A casual look at the electricity market scenario today will give us an indication that smart grid is a reality, at least with partial capabilities. However, this development necessitated the involvement of a large

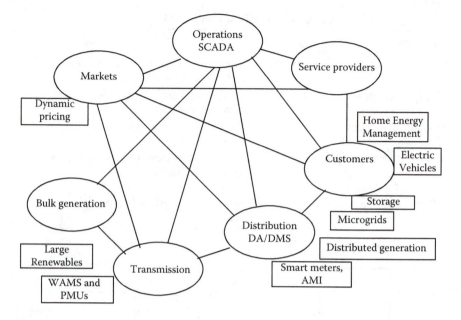

Figure 7.2 Smart grid conceptual diagram.

number of stakeholders who worked in unison with a vision to improve the grid and be prepared for the challenges ahead.

Policy makers, utilities, researchers, manufacturers, vendors, and customers are all stakeholders in this exciting development on the electrical energy horizon.

Policy makers in every country are coming out with guidelines for smart grid implementation. In 2007 in the United States, President Bush signed the Energy and Independence Security Act (EISA) under which Title XIII established federal policy on the smart grid. The National Institute of Standards and Technology (NIST), which is in the Department of Commerce, was made the overall coordinator of smart grid standards and interoperability. In 2009 in the United States, President Obama signed the American Recovery and Reinvestment Act, appropriating $4.5 billion in budget authority to modernize the electric grid. NIST received $12 million of these "stimulus funds" to fulfill its legislative mandate to coordinate smart grid standards and interoperability. The Government of India (GOI) in an effort to move toward a smarter grid implemented the accelerated power development and reforms program (APDRP) providing funding to state electricity utilities to improve the energy auditing and accounting system and hence reduce the losses by 15%. The GOI has also set up a Smart Grid Task Force, an interministerial group, to create a road map. The Smart Grid Forum is a nonprofit voluntary consortium of public and private partners with the objective of accelerating the smart grid developments in India. Similar developments are happening around the globe, and pilot projects are running at many locations. The emphasis of the policy makers is on creating advanced metering infrastructures for inclusion of customers as stakeholders, building two-way digital communication with cybersecurity, development and deployment of standards for interoperability, real-time wide-area monitoring system implementation.

Industry standards developers have been integral to development. EISA of 2007 directed the NIST to form the Smart Grid Interoperability Panel (SGIP) in 2009 as a public-private partnership. SGIP's charge was to coordinate standards work to ensure interoperability and security as the grid is modernized. After SGIP developed a strategy and a structure in 2010 to 2012, it transitioned in 2013 to SGIP 2.0, Inc., a member-funded organization that carried forth its predecessor's original mission. The transition essentially put the power industry in charge of its own destiny. SGIP 2.0, Inc. grew in 2012 from 88 members to more than 200, as various stakeholders grasped the power industry's fundamental challenges and seized an opportunity to shape the future. Work efforts include case studies, white papers, tools, and use cases addressing cybersecurity risk management, smart grid cloud computing, distributed renewable energy resources, volt-var management, dynamic pricing, electric vehicle charging, testing and certification, demand response programs, transactive retail energy applications, bidirectional weather data exchanges, global interoperability approaches, and regulator educational seminars.

Utilities are reinventing the advantages of supervisory control and data acquisition (SCADA) and automation of every component in the power sector from generation to transmission to distribution to the customers. Across the world, deregulation and restructuring have created a healthy competition and modernization of the aging infrastructure, and most of the utilities are embracing the smart grid concepts with great enthusiasm. Huge investments are being made for implementing advanced metering infrastructure (AMI) not only for industrial and commercial customers, but also for domestic customers while providing specific advantages to the utilities. Transmission utilities are installing wide-area measurement systems with phasor measurement units, phasor data concentrators, and associated data integration with the existing SCADA systems to make the transmission systems more secure and avoid blackouts.

Researchers and research laboratories have initiated the smart grid concept by designing and developing analytical tools and technologies for smart grid implementation. A consortium of academic institutions has been working toward developing software architecture for smart grid. As a National Science Foundation Industry-University Cooperative Research Center, the Power Systems Engineering Research Center (PSERC) draws on university capabilities to creatively address the challenges facing the electric power industry. Its core purpose is to empower minds to engineer the future electric energy system. Under the banner of PSERC, multiple US universities are working collaboratively toward:

- Engaging in forward-thinking about future scenarios for the industry and the challenges that might arise from them
- Conducting research for innovative solutions to these challenges using multidisciplinary research expertise in a unique multi-campus work environment
- Facilitating the interchange of ideas and collaboration among academia, industry, and government on critical industry issues
- Educating the next generation of power industry engineers

PSERC provides

- Efficient access to experienced university researchers in an array of relevant disciplines and geographically located across the United States
- Leading-edge research in cost-effective projects jointly developed by industry leaders and university experts
- High-quality education of future power engineers

The multidisciplinary expertise of PSERC's researchers includes power systems, applied mathematics, complex systems, computing, control theory, power electronics, operations research, nonlinear systems, economics, industrial organization, and public policy. PSERC partners with private and public organizations that provide integrated energy services, transmission and distribution services, power system planning, control and oversight, market management services, and public policy development.

Power system equipment *manufacturers and vendors* have contributed to the smart grid migration by creating and developing new equipment and software for implementation.

Technology companies have also jumped into the smart grid arena to utilize their expertise in the information technology (IT) sector by developing IT technology–related products suitable for the smart grid implementation, as the smart grid involves amalgamation of the IT with electrical technology. New products and applications are being developed by these companies specifically to suit the needs of the power automation industry.

7.5   Smart grid solutions

Migration to smart grid enhances the power system performance in many ways, and the major foundations of a smart grid implementation are the solutions. Traditional device and system development continues, but with

Shared Services and Applications					
Asset Optimization	Demand Optimization	Smart Meters and Comms	Distribution Optimization	Trasmission Optimization	Workforce and Engineering Design Optimization
Interoperability Framework					

Figure 7.3 Smart grid interoperability framework.

smart grid the industry has transitioned to solutions. A solution is a set of technology components integrated to successfully interoperate to address the business needs of the electric utility customers. In general, the six solutions with the strongest business cases are

1. Asset optimization
2. Demand optimization
3. Smart meter and communication
4. Distribution optimization
5. Transmission optimization
6. Workforce and engineering design optimization

Figure 7.3 gives a clearer picture of the foundations, and the following sections elaborate the functions of each foundation.

7.5.1   Asset optimization

Asset optimization includes proactive equipment maintenance via equipment condition monitoring and produces a lot of advantages for the power industry. Focused maintenance can be done on equipment, and asset optimization will definitely reduce outages and risks of failure as the assets are monitored and assessed continuously. This leads to better utilization of assets, and a utility can squeeze more capacity from existing equipment and devices. Savings are achieved due to the delayed investment in additional equipment. The details of this delayed investment are explained in Section 4.9.3.4. Asset optimization also includes use of intelligent sensors and equipment as discussed earlier and shown in Figure 7.4.

7.5.2   Demand optimization

Demand optimization is a major paradigm shift in the history of power distribution due to customer involvement. The main feature of demand

Shared Services and Applications		
Intelligent Sensors	Monitoring and Diagnostics	Desision Engine

Figure.7.4 Asset optimization.

Shared Services and Applications		
DR Mgmt. Software (DRMS)	Comms. (ISP)	In-Home Enabling Technology
Interoperability Framework		

Figure 7.5 Demand optimization.

optimization is the peak consumption reduction by consumers, thereby reducing the peak load and hence the generation at the utility. This is achieved through distribution management systems, in-home enabling technologies, such as smart meter, and the associated communication infrastructure, as shown in Figure 7.5.

For example, the United States has an installed capacity of 1,000,000 MW, and if 20% of this capacity is used only 5% of the time, this essentially means that 200,000 MW capacity generation, transmission, and distribution resources worth $300 billion are utilized for only 5% of the time. It may be worthwhile to see how these assets can be better utilized if we understand demand response by involving customers. This can be achieved by incentivizing the customers for opting for demand response while implementing automated load cuts from the substation along with additional incentives.

Thus demand optimization results in delay or avoidance of additional investment; thus, better utilization of the existing infrastructure is achieved. It also encourages customer empowerment, increases satisfaction and loyalty of the customers, and allows customers to save money by reducing electricity bills.

7.5.3   Distribution optimization

Distribution optimization revolves around the automation of the distribution system, as discussed in Chapter 6, which involves automating, the distribution substations, feeders, and the customers. Although cost

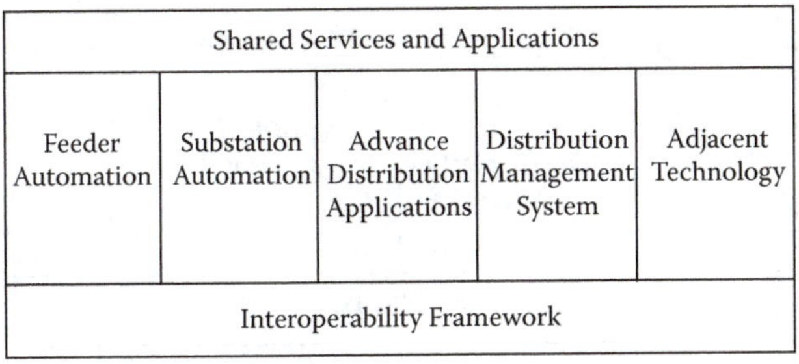

Shared Services and Applications				
Feeder Automation	Substation Automation	Advance Distribution Applications	Distribution Management System	Adjacent Technology
Interoperability Framework				

Figure 7.6 Distribution optimization.

intensive, the distribution automation brings added advantages to the utility as it opens the pathway to the smart grid implementation and all the associated functionalities.

Another important feature of distribution optimization is the renewable integration at the feeder level and at the customer level. This integration creates additional advantages like reduction in peak demand, deferred capital investment, satisfied customers, and so on. However, it causes integration and operational issues for the distribution system, unpredictable output of the renewable resources, storage problems, and finally the structure development. Distribution optimization will ensure less energy waste, with higher profit margins by reducing the losses in the distribution systems. Figure 7.6 shows the components of a distribution optimization solution.

Plugged hybrid vehicle integration is another aspect of distribution optimization. With more companies manufacturing electric vehicles, it is imperative for the distribution utility to utilize adequate charging facilities and infrastructure to cope with the growth.

Thus, distribution optimization allows cleaner, greener generation, which implies emission reduction with improved efficiency and reliability.

7.5.4   Smart meter and communications

Smart meter is the brain of customer automation and in a way the implementation of demand optimization. Smart meter with well-defined functionalities at the customer premises enables both customers and the utility to reap the benefits of automation. A two-way communication infrastructure for the smart meter interaction with the utility is a challenge, because as in distribution systems, these have to be developed from scratch. Utilities are investing in this segment so that the benefits of smart grid can be reaped completely. Enabling technologies for network connectivity, consumer enablement, demand optimization, and improved

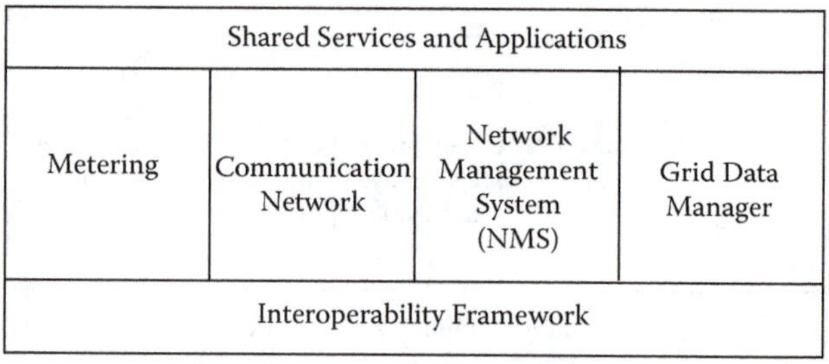

Shared Services and Applications			
Metering	Communication Network	Network Management System (NMS)	Grid Data Manager
Interoperability Framework			

Figure 7.7 Smart meter and communication solutions.

operations are a few features of smart meter and communication optimization. Figure 7.7 gives the smart meter and communication infrastructure solution architecture.

7.5.5 *Transmission optimization*

Transmission optimization involves wide-area monitoring protection and control, and it improves the reliability and efficiency of the transmission. Earlier the state estimation and other related transmission SCADA applications were dependent on the 2 s status and 10 s analog data acquired by the SCADA system. With the deployment of phasor measurement units and phasor data concentrators, the state estimation and related applications produce more reliable and faster data, and the operators are better equipped to deal with contingencies. Great improvement in reliability and efficiency of the system is achieved by the WAMS implementation. Figure 7.8 shows the components of transmission optimization solutions.

Another interesting feature of transmission optimization is the integration of large renewable energy sources to the grid, such as wind farms and solar farms. This leads to cleaner generation although it brings in a set of problems and challenges in operating the grid, mainly due to the uncertainty involved in the renewable energy prediction. Transmission optimization had been dealt with in detail in Chapter 5.

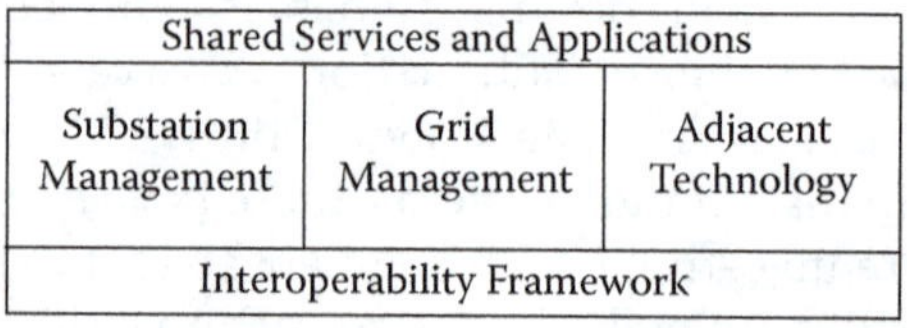

Shared Services and Applications		
Substation Management	Grid Management	Adjacent Technology
Interoperability Framework		

Figure 7.8 Transmission optimization.

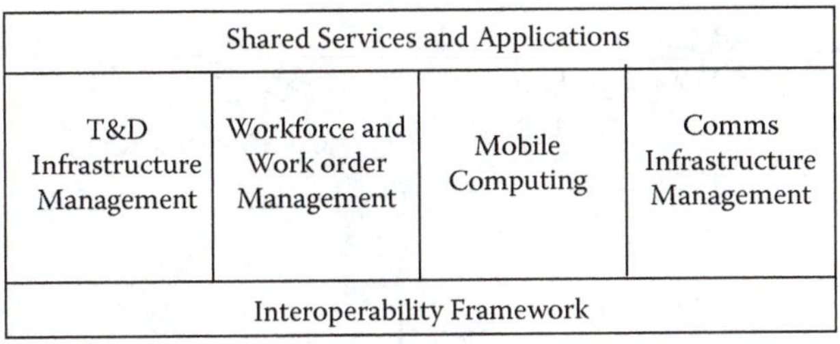

Shared Services and Applications			
T&D Infrastructure Management	Workforce and Work order Management	Mobile Computing	Comms Infrastructure Management
Interoperability Framework			

Figure 7.9 Workforce and engineering optimization.

7.5.6   Workforce and engineering optimization [4]

A rapidly evolving workforce is reshaping the risk profiles of power and utilities organizations, posing challenges to their traditional control and compliance capabilities. A more systematic approach to capturing and keeping core know-how and new ways of transmitting that knowledge to a younger generation are necessary. Approximately 40% of current employees and 60% of current executives are eligible to retire in 5 years in the USA for instance. Workforce-enabling technologies such as automated workforce deployment and field force automation have led to increased workforce productivity and work satisfaction. As discussed in Chapter 6, many utilities are engaged in updating and integrating the functionalities in a distribution control room such as DMS, OMS, GIS, AMI, and so on, so that the dispatcher and other workforce get maximum information and effective tools for data usage. Data warehousing to help the different departments of the electric utility has already been discussed.

Engineering optimization involves improved modeling and system design capabilities, so that cost-effective system design and modeling can be done with ease. It is seen that workforce optimization can save up to 30% labor cost in a well-motivated, automated utility. Figure 7.9 gives the workforce and engineering optimization architecture.

7.5.7   Smart grid road map

Figure 7.10 shows the smart grid road map with all the six solutions depicted, which gives a clear understanding of the path ahead.

7.6   Smart distribution

Governmental policies covering reduction in pollution emissions, climate change, and incentives for renewable generation are providing electric utilities enough reasons to be involved in major revamping of the

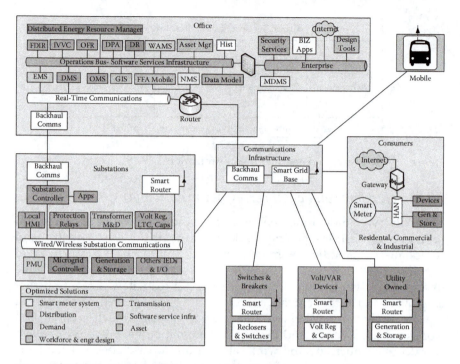

Figure 7.10 (See color insert.) Smart grid road map.

distribution infrastructure. The needs and requirements of the customer are also changing faster, and the utility has to provide innovative solutions with quality power to the customers, both industrial and domestic. The distribution system needs to be redesigned for high levels of distributed energy resources, creating a smarter and more flexible system for improved reliability, high penetration of renewable sources, dynamic islanding, distributed control, and increased generation efficiencies through the use of waste heat. The following sections will elaborate the smart distribution system building blocks. It may be noted that the smart distribution system becomes a reality when systems are automated and ready for smart operation.

7.6.1   Demand-side management and demand response [3,5]

Electric energy demand-supply balance is the basis of stable operation of a power system. The large generation, bulk transmission and distribution systems work in unison to achieve this balance at all times, but bulk energy storage remains a challenge. However, the inclusion of customers—the demand side of the distribution system—is attracting a lot of attention as this can help the utility to streamline the operations with a host of advantages. Demand-side management attempts to improve energy efficiency at the customer side of the electricity distribution.

Demand-side management is a solution to equip the power system to meet the ever-increasing demand of electricity around the globe, where the traditional generation and transmission systems are not able to handle additional load requirements. Intelligent demand-side management is made possible due to the development of two-way communication facilities between the distribution utility and the customer, thus paving the way for customer involvement. Intelligent DSM helps utilities to stretch the limits of the generation, transmission, and distribution, so as to accommodate the demands of the customers, by managing the load on the system. Distributed generation is an added advantage here, as the local generation to supply the load is a way of decongesting the transmission and primary distribution systems.

Figure 7.11 shows the load management scenarios used in different time frames and the impacts on the process quality at the customer end. Until recently, the entire activity was "utility driven"; however, in the smart grid perspective, the shift is toward "customer driven."

Hence, demand-side management can be divided into four categories depending on the timing and impact of the measures used for load reduction, as given in Figure 7.11:

- Energy efficiency (EE)
- Time of use (TOU)
- Demand response (DR)
- Spinning reserve (SR)

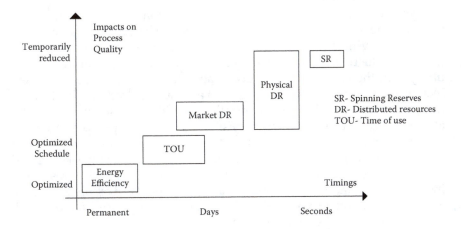

Figure 7.11 Demand-side management time frames [5].

7.6.1.1   Energy efficiency (EE)

The best way to curb consumption of electricity is by making the end customer devices more efficient to optimize energy use. Governments all over the world are working toward legislation to help customers to buy energy-efficient appliances such as refrigerators, lighting, and efficient industrial equipment. However, this requires the customer to buy more efficient devices or replace the old ones and invest in better appliances. All other techniques discussed only shift the peak load, but there is no reduction in the energy use.

7.6.1.2   Time of use (TOU)

Here the electricity tariffs are different for different periods of time in a day when the utility expects peak loads. Typically the morning hours and evening hours for domestic customers are when the use of electricity peaks. However, if the prices are higher during these hours, customers can reduce consumption by shifting some of the activities like pumping water, washing clothes, and so forth, to off-peak hours. Section 7.6.1.4 has demonstrated the fact that in most of the countries, the peak 15% to 20% of the demand exists for 0.5% to 5% of the time in a day. This presents a perfect scenario for the utilities to incentivize the customers to stagger the peak loads and pay lower bills. This has been successfully implemented in many countries around the globe, and utilities have delayed investment for additional generation and transmission facilities to serve the peak load at specific times. Here the utility has no direct control on the demand, and conscientious customers will make an effort to stagger the load so that they get reduced power bills.

7.6.1.3   Demand response (DR)

Demand response allows customers more freedom to choose what load they want to shed and how much they want to shed for incentives. One way of achieving the load reduction is by having a controller in which the load reduction is predetermined by stochastic programming by the utility, and the customer noncritical loads (predecided by the customer, like air conditioners, heating units, etc.) are switched off for short durations by the utility to reduce the overall load at the substation. This could also come into play in an emergency situation like maintenance or outage of equipment.

Demand response can be a market issue under which market prices decide the demand response mechanism or can be a physical response involving grid management and emergency control. For optimal grid operation, a good mixture of market demand response and physical demand response is generally adopted.

In a distribution system, *spinning reserve* (SR) is the other end of the demand-side management spectrum. Loads can act as virtual spinning reserves (negative) when they correlate their power consumption to the state of the grid; as the frequency dips, the power consumption of the loads reduces. This can happen in an autonomous way in a primary control and in a coordinated way in a secondary control.

Another way of classifying demand response is as incentive-based DR and as time-based rates DR:

Incentive-based DR
- Direct load control (DLC): utility or grid operator has free access to customer processes
- Interruptible or curtailable rates: customers have special contracts with limited sheds
- Emergency demand response programs: voluntary response to emergency signals
- Capacity market programs: customers guarantee to reduce usage when grid is in need
- Demand bidding programs: customers can bid for curtailing at attractive prices

Time-based rates DR
- Time-of-use rates: static price schedule is applied
- Critical peak pricing: a less predetermined variant of TOU
- Real-time pricing (RTP): wholesale market prices are applied to end customers

7.6.1.4   Peak load on the system: Case study

In a typical utility system with an installed capacity or peak load of 19,140 MW, the peak load exceeded 16,000 MW for only 5% of the time, and hence 3140 MW was on the system for only 5% of the time, as shown in Figure 7.12 (Dominion Power, Virginia). Of this, peak demand can be redistributed on the load curve by appropriate means, by customer involvement, and effectively reduces the peak load by 16.5% (3140 MW).

The above observation regarding the peak load duration is true in most countries. As mentioned earlier, in the United States 20% of the load is needed for only 5% of the time, whereas in Australia and Egypt, 15% of the load is needed in less than 1% of the time. In Saudi Arabia, 15% of the load happens tless than 0.5% of the time. The above data give excellent insights into the opportunities for reducing capital investment through demand optimization and customer participation. The shaving of the peak load will release additional power for the new loads coming up and reduce investment for adding additional generation capacity (FERC).

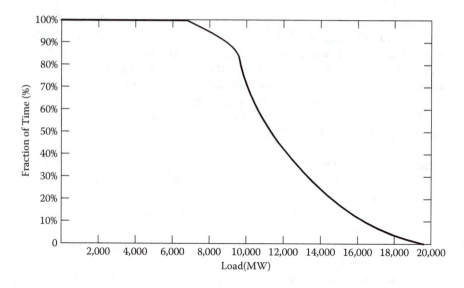

Figure 7.12 Load versus fraction of time illustration for Dominion Power, Richmond, Virginia.

This is just an example of the tremendous potential demand response will provide if properly implemented and executed.

7.6.2 Distributed energy resource and energy storage [4,6,7,8]

In the smart grid scenario, the smartness demands sources of electric power that are not generally connected to the bulk transmission system but are near the load in a distribution system. Such smaller sources that are located and interconnected at the local customer or distribution level are named *distributed energy resources* (DERs). Typically, the individual DER unit ratings are less than 10 MVA and include both fossil fuel and renewable generation and energy storage technologies.

DERs provide a variety of advantages to the utility, as they supply energy at the local level and reduce the central generation capacity addition, some of which are listed below:

1. Relieving the transmission and distribution system of the loading; transmission and distribution capacity deferral, thus saving money and relieving congestion
2. Central generation capacity deferral, thus less investment for the utility
3. Avoiding T&D losses, thus energy saved
4. Improved power quality due to the local power supply
5. Better voltage and var control
6. Reduced emissions and greener environment

However, the integration of DERs into the distribution system introduces a multitude of problems, as this concept envisages two-way power flow, and the normal distribution systems are not designed accordingly. The operation of the distribution system will become complex, the protection system design will be more challenging, and the control strategies will be complicated.

DERs include distributed generation, local electricity generation, as well as storage devices for electrical energy storage.

7.6.2.1   Distributed generation (DG)

The CIGRE WG 37-23 [2003] defines the DG unit as a generation unit that is not centrally planned, not centrally dispatched, usually connected to the distribution network, and smaller than 50 to 100 MW. The IEEE Standard 1547 [2003] defines the DG as generation of electricity by facilities of size usually 10 MW or less, so as to allow interconnection at nearly any point in the power system.

Hence, DG includes the variety of small generation units ranging from solar arrays, small hydro, wind turbines, geothermal power, micro turbines, fuel cells, diesel engines, tidal power, and wave power, to name a few. The DGs are now an integral part of most of the distribution systems across the world and are owned and operated by utilities and customers alike.

7.6.2.2   Energy storage [9,10,11]

Energy storage can be defined as the conversion of electrical energy from a power network into a form in which it can be stored until converted back to electrical energy.

Electrical energy has the special characteristic that the generated power has to match the load demand and that makes it a very dynamic system. Also, the load centers are generally located at far distances from the bulk generating units. This necessitates large transmission systems, and the outage of lines creates severe power shortages. Congestion of a few power lines is also caused during peak load hours. Generally peak load is met by operating costly generating units such as oil and gas power plants; however, there exists a huge potential if the electrical energy generated by the base load plants during off-peak hours (especially nights) can be stored in some form and used to meet the peak loads at a later stage.

The widespread use of renewable energy sources such as solar and wind power also brings energy storage to the forefront, as these sources are inherently intermittent in nature, and an energy storage facility can greatly enhance the potential of these sources.

To maintain the voltage and frequency of the supply within permissible limits, energy storage can play a major role at substations. This energy storage can act as a source of energy in a distribution system and can help mitigate a number of problems such as supply of peak load, power quality improvement, stability improvement, power smoothing, voltage regulation, and related issues.

In home energy management systems, energy storage plays a major role in smoothing the problems of power quality, grid failure, and plugged hybrid vehicle integration. Energy storage can be utilized effectively to help utilities and consumers to deal with inherent problems in providing an uninterrupted supply of quality power.

Figure 7.13 gives a comprehensive view of the electrical energy storage at different stages in the distribution system and the related effect on the distribution system.

We now discuss the classification of electrical energy storage. An electrical energy storage system can be classified according to the form of energy used. Mechanical, electrical, thermal, chemical, and electrochemical forms of energy storage are available:

1. Mechanical
   a. Pumped storage system
   b. Compressed air
   c. Flywheel storage
2. Electrical
   a. Double-layer capacitor
   b. Super-conducting magnetic coil
3. Thermal
   a. Sensible heat storage
4. Chemical
   a. Hydrogen (fuel cell, electrolyzer, etc.)
5. Electro-chemical
   a. Secondary batteries (lead-acid, Ni Cd, NiMh/Li, NaS)
   b. Flow batteries (redox flow/hybrid flow)

In Figure 7.14 ([11], The Fraunhofer Institute for Solar Energy Systems, Frankfurt, Germany, compared the energy content, rated power, and discharge time of the different storage devices which gives a clear understanding of the usage of each device for mitigating different problems faced by a utility as discussed earlier.

Pumped hydro and hydrogen storage systems having large discharge times of a few hours can be built for maximum capacity of a few megawatts, whereas the electrical storage systems like double-layer capacitors and superconducting magnetic coils are realized for fast discharge times of a few seconds and lower capacity. The capacity limitations are mainly due to economic reasons, as the systems are modular and may be expanded once the technologies become cheaper.

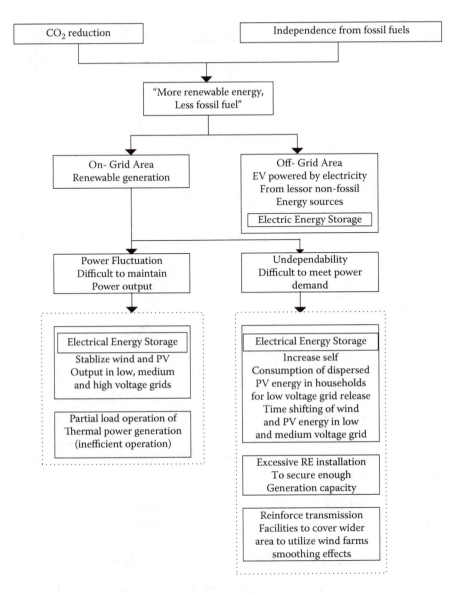

Figure.7.13 Energy storage flow diagram [11].

The current deployment of the electrical energy storage across the world also presents an interesting study. Over 99% of the electrical energy storage is hydro with an installed capacity of 127,000 MW across the world. Compressed air is around 440 MW, sodium sulfur battery is around 316 MW, fly wheel is around 25 MW, with capacities for lithium

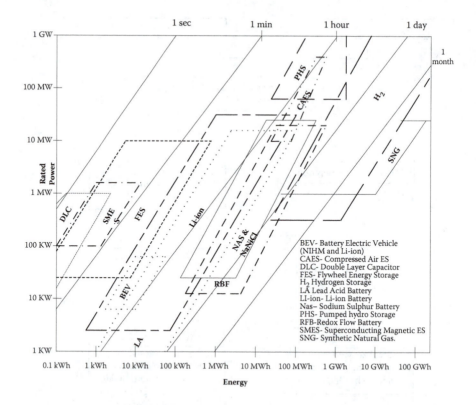

Figure 7.14 Energy output, capacity, and release duration of energy storage devices [11].

ion (70 MW), lead acid (35 MW), nickel cadmium (27 MW), and redox flow (3 MW) batteries.

Thus energy storage gives an option for the electrical system to mitigate many problems created by the dynamic power systems to maintain load generation balance and to ensure power quality.

7.6.3   *Advanced metering infrastructure (AMI) [12,13,16]*

The smart grid concept revolves around motivating and including customers in the grid management in various ways. The two-way communication between the utility and the customer, seen as the paradigm shift in the way customers are engaged by the utility, is achieved by the deployment of the AMI.

The AMI systems measure, collect, and analyze energy usage, from advanced devices such as electricity meters, gas meters, and water meters through various communication media. However, for a power utility, the AMI network provides the communication link between the customer and

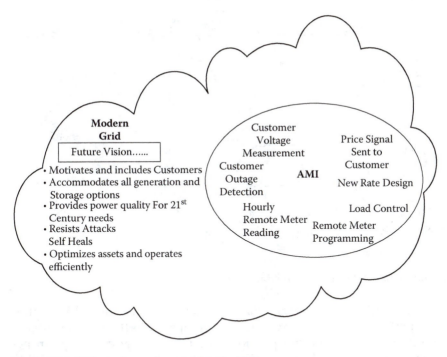

Figure 7.15 Vision of a modern grid with AMI.

the utility and provides measurements and system observability. Thus, AMI is not a single technology, but rather an integration of many technologies that provides an intelligent connection between consumers and system operators. AMI gives consumers the information they need to make intelligent decisions, the ability to execute those decisions, and a variety of choices leading to substantial benefits they do not currently enjoy.

Through the integration of multiple technologies such as smart metering, home area networks (HANs), integrated communications, data management applications, and standardized software interfaces with existing utility operations and asset management processes, AMI provides an essential link between the grid, consumers and their loads, generation, and storage resources. Such a link is a fundamental requirement of a modern grid. Figure 7.15 depicts the vision of the modern grid and AMI is the first step toward grid modernization.

7.6.3.1 Components of AMI
AMI is composed of components that have been integrated to perform as a single platform to provide inputs to other automation systems such as distribution automation, outage management, and customer services. Figure 7.16 shows the AMI structure and the data flow and interface.

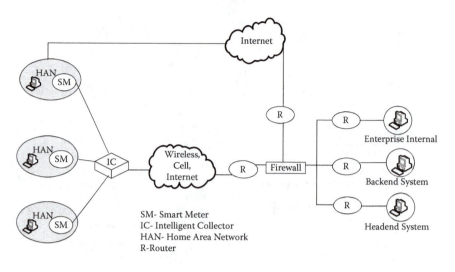

Figure 7.16 AMI structure and data flow.

7.6.3.1.1 Smart meters. Smart meters are programmable devices that perform a variety of functions as compared to the traditional electromechanical meters that provided the total energy consumption per month per customer. The new digital meters with two-way communication between the meters and a remote system are capable of performing a large number of functionalities as follows:

- Provide real-time information of electricity usage by the users
- Communicate the meter readings to a remote station
- Provide power quality assessment by storing waveform information
- Accept the commands from the remote station to turn off specific appliances at the household
- Provide flexible tariff information to the consumers
- Provide digital output connections for interfacing the devices for remote control
- Perform remote turn off/on operations
- Notify of loss of power
- Provide tamper and energy theft detection
- Communicate with other intelligent devices in the home (home automation systems)

7.6.3.1.2 Intelligent collectors (ICs) The intelligent collectors establish a secure connection with a specific group of smart meters and report energy usage periodically to the head end application at the utility. The ICs are paired securely to the head end application at the local level. The control commands and other information received from the head end application are forwarded by the ICs to the smart meters at the customer premises. Thus the two-way communication as envisaged by the smart grid is established between the customer and the utility.

7.6.3.1.3   AMI head end   The AMI head end polls the group of meters through the intelligent collectors as per the predefined intervals for collecting the energy readings and sends the information to the meter data management system (MDMS). The AMI head end system initiates the commands to the customers in general and the specific appliances in particular at the customer premises to facilitate the demand response mechanism. The AMI head end system acts as a central commanding station for performing the control operations apart from polling the meter readings.

7.6.3.1.3   Meter data management system (MDMS)   The MDMS located at the distribution utility level stores the meter data and performs a number of functions with the collected data. The MDMS validates the data, analyzes the data, and does the necessary editing to convert it into bill format. The bills are then shared with the customer information system (CIS) and other service providers.

7.6.3.1.4   Communication infrastructure   This is a major component. The whole concept of AMI depends heavily on the communication capability available, and this poses the greatest challenge for smart meter implementation. The communication happens between the smart meter and the intelligent collectors (last mile connectivity) and between the intelligent collectors and the head end located at the substation or local control center. Ensuring cheaper, secure communication without intrusions is a challenge in the implementation. Table 7.2 provides a comparison of automated meter reading (AMR) and advanced metering infrastructure (AMI).

7.6.3.2   AMI integration with DA, DMS, and OMS [15]

The many benefits of distribution automation (DA)—visibility, fault detection and isolation, energy efficiency, and asset management—are creating a "second wave" of smart grid investments and integrations, following the widespread adoption of AMIs. Currently, in fact, the business case for DA is better than for any other single system in the phased steps of grid modernization.

In those phased steps, typically AMI comes first, followed by DA. In fact, DA relies on AMI's end-of-line sensors, a.k.a. "smart" or interval meters, to enable its benefits. The hitch in this picture lies in the fact that the fundamental AMI system must accommodate DA functionality.

Utilities that adopted AMI as an end in itself, without a well-considered technology road map, may be rudely surprised to learn that their AMI choices of a few years back were not made with future integrations in

Table 7.2 AMR versus AMI

	Automatic meter reading (AMR)	Advanced metering infrastructure (AMI)
Meters	• Electronic	• Electronic with LAN, HAN, load profile, and disconnect
Data collection	• Monthly	• Remote using LAN communication, more frequent
Data recording	• Cumulative kWh	• Half hourly so that dynamic prices can be applied
Business opportunities	• Monthly consumption-based billing	• Customer payment option • Pricing options • Utility operations • Demand response • Emergency response
Key business processes	• Billing • Customer information system	• Billing • Customer information system • Customer data display • Outage management • Emergency demand response
Customer's participation	• None	• Demand response programs • In-home display, cost and environmental consciousness
Additional devices	• None	• Smart thermostats/compressor • Utility-controlled customer load
Outages	• Customer phone calls	• Automatic detection • Verification of restoration at individual home level

mind. We need an industry standard that details the architecture of communications infrastructure within the meter so it can send "last gasps" to the outage management system (OMS), voltage data to the distribution management system (DMS), and serve other functions to systems other than AMI. But because there is no such industry standard with respect to AMI systems, the endeavor remains very much a *caveat emptor* situation.

Some AMI systems simply do not lend themselves to DA integration and may require replacement or a laborious, expensive, inefficient workaround. For utilities that have not yet embarked on an AMI implementation, looking ahead to future systems integration can avoid duplicative efforts and costly mistakes. In fact, a successful DA integration with AMI unlocks the value in both systems. In addition, the creation of a technology road map and adoption of these and other technologies should drive organizational change toward a more holistic approach to smart grid. De-siloing will bring efficiencies and further unlock the value in technology adoption, something that regulators will increasingly demand as they scrutinize cost recovery and rate cases.

7.6.3.3   *The market and the business case [15]*

Of the approximately 48,000 distribution substations in the United States, fewer than half have any sort of automation. Substations with some automation and those without automation typically connect to feeders with no automation or monitoring whatsoever. Today, very few distribution feeders send any kind of real-time information upstream. This creates large areas of, shall we say, "unobservability." We just do not know what is happening on the system.

The imperative to learn what occurs on a system will only grow stronger, and the need to support a variety of data streams from the field and route them efficiently is only going to grow exponentially in the near future. As more distributed renewable energy is integrated into the grid, and as a utility copes with two-way power flows, the utility will face new safety and protection challenges. Add to that the additional, two-way data flows that will accompany dynamic pricing and the interaction of that signal with a home energy management system. When the peak price of electricity moves customers to shed load, the utility will want to understand precisely how much load is being shed, individually and in aggregate. Ideally, these data flows, like those for AMI and DA, would use the same communication network: the AMI support infrastructure.

All of these considerations are driving utilities to implement a DMS to manage complexity. But the DMS is only as good as the information coming from the field. These factors explain why distribution automation or distribution optimization, if you will, currently represents the most cost-effective step and the best business case of all smart grid solutions.

The substations and feeders without automation in the United States pose an enormous challenge—not coincidentally, a huge addressable market for vendors and a cost-effective route with big operational and organizational payoffs for utilities. The thinking outlined here is designed to assist in making the best products and the most cost-effective investments.

7.6.3.3.1   *First, break down the walls*   The holistic approach to smart grid, in particular, and grid modernization, in general, requires a strong dose of executive leadership due to entrenched interests that have persisted in the power utility culture. For instance, AMI implementations typically fall under the purview of a metering group inside the utility, while DA is under a distribution engineering group in operations. The two systems share a need for service territory-wide communications systems. Too often, a siloed utility builds two systems, side by side, when a single, well-vetted system could serve both purposes. That results in redundant

efforts and expenses and two separate data streams that would serve the organization far better if they were integrated.

The simplest way to avoid this misstep is to have executive leadership, sometimes aided by a third party, bring together the metering group and the distribution engineering group to jointly determine their mutual, functional requirements for a common communication network. And do not stop at accommodating DA functionality, because as mentioned, demands on that network will only grow with time.

Cooperation leads to a stronger business case for both systems in this example. A general rule of thumb for a technology road map and resulting utility investments is to develop them with a horizontal organizational structure that results in cost-effective investments and integration-friendly systems. As this becomes a more widely recognized best practice in the smart grid era, regulators will come to expect this approach and conceivably may base decisions on whether it is being implemented.

7.6.3.3.2 Integrating the acronyms Many AMI technologies are designed for meter-related data output only—those 15 min interval readings that flow upstream to the network management system (NMS) that manages the communication network aspect of AMI and also feeds the data to the meter data management system (MDMS). The MDMS stores and feeds the data to applications, such as generating customer bills, analyzing usage patterns and so forth.

The interval meter's "last gasp" when an outage occurs is not metering information; that signal needs to be routed to the OMS where it can be analyzed to determine the cause and extent of the outage. Some AMI systems cannot split off that last gasp to the OMS. Similarly, in another distribution automation function—voltage data coming back from the end-of-line sensor—the meter, needs to be routed to the DMS to ensure that the utility is achieving the 114 V to 126 V ANSI standard at the customer premise. That is not easily accomplished with some AMI systems. Note that one does not need the voltage readings from every meter, just those at strategic points at the ends of selected feeders.

An AMI system is the glue between the meter and the utility. Functionality in the meter needs to be matched to functionality in the supporting systems, the "infrastructure" in AMI. That means the communication network. Thus, an AMI system needs a certain flexibility to integrate properly with DA functions such as routing meters' last gasps to the OMS and steering voltage information to the DMS.

For utilities that have installed AMI, this underscores the need to evaluate the underlying systems with DA integration in mind. A utility may have had the foresight to develop a carefully thought-through road map and be in a good position to reap the benefits of DA. If that foresight was lacking, the consequences can be laborious and expensive. It is technically true that AMI data can be routed through the NMS and the MDMS to reach the OMS and DMS, but that is a cumbersome route that challenges bandwidth and latency. A well-architected system would avoid that scenario.

Further, as meters gain functionality, they may well be upgraded or swapped out for more advanced ones. What a utility wants to avoid is replacing the underlying infrastructure—again, the "I" in AMI.

The AMI system needs to have enough flexibility to support the metering information going to the NMS and MDMS, but also support other data outputs on the smart meter and be able to route them to other systems. Routing last gasps to the OMS and sending voltage data to the DMS were mentioned, but as time goes on, deriving value from more functionality in the meter requires having the capability to route those data streams to other systems and destinations over a common communication infrastructure.

7.6.3.3.3 Vetting the DMS The DMS contains the network model manager, which is a critical piece of software. Utilities would be wise to look closely at this functionality during the procurement process. The DMS must interface with the utility's geographic information system (GIS), so it is imperative to know whether the DMS in question will, in fact, integrate easily with the utility's GIS. The DMS must know what data to pull from the GIS, how that information is stored, and how to retrieve the needed data for building the network model.

A DMS that works well with a GIS is important because, as the data in the GIS changes, incremental updates inform the network model in the DMS and keep it up to date. The OMS also has a network model for outage analysis that depends on the GIS as well.

7.6.3.3.4 The network model Think of the network model as two major sets of information. One set is the power system connectivity information, which includes the electrical characteristics of grid assets. For instance, that includes transformers, the model of each transformer, and its connection information—is it YY, is it ΔΔ, is it ΔY grounded? Power system connectivity information also includes the branches, nodes, and capacitor banks connected on the distribution feeders to ground.

The second set of information in the network model is the real-time information about the network, the operational information—the voltage, current, real and reactive power flows, statuses of switches and circuit breakers, and so forth.

7.6.3.3.5 DA functions, up close and personal When we talk about DA, we mean three primary functions: improving reliability with fault detection, isolation and restoration (FDIR) for optimal feeder reconfiguration;

reducing losses with VAR control; and managing load or demand with voltage control. (Voltage is directly proportional to load, so when we control voltage, we control load. VAR is a reactive power, directly proportional to losses.)

Today, with DA, the utility can combine voltage and volt-ampere reactive control with integrated volt-var control (IVVC). In fact, a DMS optimizes these applications. But to do so, the DMS requires real-time information about what is happening on the distribution system downstream of the substation.

To assess whether an AMI system will support DA functionality, one needs to weigh the response requirements of the DA applications. Three metrics must be assessed: speed, bandwidth, and latency. For instance, FDIR requires a 2 to 3 s response for rapid switching. (Those are SCADA-level speeds.) Capacitor controls require about 30 to 60 s. Many AMI systems are designed to support only 15 min interval reads, yet intelligent electronic devices (IEDs) often need to send megabytes of data upstream at one time, requiring speed, bandwidth, and low latency.

Here are some questions to ask. Can your utility countenance delay in enactment of operational commands? Are hundreds of milliseconds of latency tolerable? There are other considerations. Cybersecurity practices such as encryption affect the performance of data communications, increasing latency. Seeking the 200 ms latency one is accustomed to while adding "overhead" in the form of security measures may not be realistic.

When the utility adds sensors at both substations and feeders, much more information heads upstream. That has an impact on the system's ability to meet the response requirements of DA applications, in terms of speed, bandwidth, and latency.

7.6.3.3.6 Avoiding the Abyss and Stranded Assets The utility needs to ask hard questions of its vendors to avoid the downside described here—making a short-sighted investment in AMI.

Is there a migration path with your vendor? What is on that path? An easy "board swap"? A more difficult, more expensive "box swap"? Maybe there is no swap; maybe it just does not exist. If there is no path forward, will that result in a stranded asset? In SCADA procurement, for instance, if a vendor said it would be supported by a top-end X-brand server in the family of servers, if you need greater computing capacity, the system has no way to grow if it is based on the current top-end equipment.

Thinking through your technology road map with a good understanding of succeeding systems' functional requirements will lead to better results and more cost-effective investments. Hopefully, this exposition of the technology challenge in integrating DA with AMI and the importance of the road map contributes to better choices for stakeholders.

7.6.4 Smart homes with home energy management systems (HEMs) [16,17]

The previous section explained the importance of the smart meter in smart grid implementation. However, to utilize the full potential of the smart meter, the implementation of a home energy management (HEM) system is essential. The HEM improves the communication between the consumer and the utility and helps in implementation of the demand response, energy saving, and reduction in electricity bills of the customers.

HEM monitors and controls the energy consumption of a household and minimizes the energy leakage to make it a smart home. The monitoring quantities could include the electric power, gas, and water supply to the household as all the meters could be monitored and controlled by HEM.

The architecture of an HEM system will thus include smart sensors, measuring devices, smart appliances, a home area network, and the associated programs to run the HEM systems, as shown in Figure 7.17.

The components of HEMs include the following:

- *Measuring Devices*: Measuring devices are significant components of the HEM which measures the energy consumption of the house. Electricity, gas, and water meters are the measuring devices that inform the HEM about the energy consumption pattern of the household. The smart power meter with a variety of capabilities as discussed earlier gives a variety of measurements at definite time intervals to the utility and to the HEM.

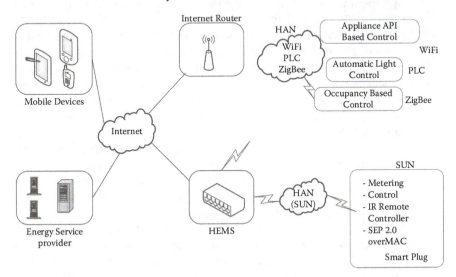

Figure 7.17 HEM systems and the overall setup.

- *Sensors*: To make a household energy efficient, and for better control of devices and appliances, a variety of sensors can be employed. A passive infrared motion sensor can be used to turn a light off and on when someone enters a room, and at the same time can be used for a burglar alarm when an unwanted movement happens. Occupancy sensors are more sensitive and can detect the presence of a person in a room. Similarly, a temperature sensor combined with a humidity sensor can accurately control the air conditioning of a room. Carbon monoxide sensors and fire alarm sensors are used for detecting fire, by optical, ionization or heat detection. There are close to 30 types including sensors for flood and moisture, proximity, contact sensors, glass break sensors, and even driveway probes to inform the HEM about a vehicle arrival, to open the gates, and other applications. RFID (radio frequency iIdentification) sensors also have application in home automation systems.
- *Smart Appliances*: Smart appliances will have the required intelligence, communication, and control facility to assist the HEM to implement the requisite decisions to make them energy efficient. Smart appliances include washing machines, dryers, dish washers, refrigerators, freezers, and ovens that can be controlled remotely with a mobile device.
- *Home Area Network (HAN)*: A home area network connects all the components of the HEM system, including the sensors, measuring devices, smart appliances, and any displays into a network for implementation by transferring the monitoring and control data as required. There are different technologies used in building the HAN backbone, depending on the communication technology and protocol used. Efforts are ongoing to standardize the technologies used, and the three technologies include (1) Zigbee wireless standards that connect the widest range of home devices, to work together with the control facility; (2) using the power line wiring in the network with smart plugs that will have specific IP addresses and can be monitored and controlled by the HEMs; (3) using the Z wave open standard for wireless which will enable the compatible devices to communicate and build an effective HAN. However, integrating various technologies for a homogeneous HAN is still a challenge as interoperability is an issue and also the security and privacy of the customer information must be ensured.
- *EMS Software*: The energy management system software for HEM will monitor the inputs, analyze the data, and with inputs from the utility and consumer (owner) devise a suitable operating plan that will be energy efficient and environment friendly. The HEM thus implements the operating plan, which can be set automatically or

manually by the owner and can be accessed from anywhere via a mobile device.

- *Distributed Energy Resource Integration*: The customers of electricity are encouraged to generate their own power at home, and solar panels are being installed at residences. There are other forms of energy generation being utilized like biogas, biomass, and solar-thermal for heating. The integration of these resources with the electric supply to a house will be monitored and controlled by the HEMs. Many houses also use batteries for storing the energy for use during a power outage, and these are also integrated by the HEMs. The plugged hybrid electric vehicles that require charging and can supply power to the house with a fully charged battery complete the list of DERs in a household.

The HEM systems are getting more versatile with entertainment systems and healthcare devices linked to the HEM. The medical alert system can be added to the HEM which will be helpful for persons with medical conditions like diabetes, arthritis, or heart disease and physically challenged persons. The medical alert system can alert the person about medication and automatically contact family members for medical help. HEM is also being extended to home office systems.

7.6.5   Plugged hybrid electric vehicles

Electric vehicles (EVs) are fast becoming popular in many parts of the world due to the many advantages they provide, especially where the fossil fuels are getting depleted while they continue to cause environmental pollution. Fully electric vehicles are on the market, along with hybrid vehicles that combine electric power and internal combustion engines. There are also fuel cell–based EVs.

The electric vehicle battery has high energy storage capability, and the charging requirements will stress the electric grid. With manufacturers of EVs planning to launch large numbers into the market in the near future, the design and operation of the smart distribution systems will be greatly influenced by the PHEVs. There will be a sharp increase in the energy demand, changes in the demand pattern, substation transformer loading, and many other associated problems. The battery in the EV is also a source of energy as it can absorb and store energy and release it to the grid or household when required when the vehicle is parked. This is the concept of vehicle to grid (V2G), which leads to many ancillary services like spinning reserve and peak power. The EVs will have to be equipped with electronic interfaces for grid connection at any moment and anywhere with distinct IP addresses which allow controlled energy exchange, metering capabilities, and a bidirectional communication interface to communicate to an aggregator entity, which will manage the large number of EVs.

A large deployment of EVs will involve [18]

- Evaluation of the impacts that battery charging may have on system operation
- Identification of adequate operational management and control strategies regarding batteries' charging periods
- Identification of the best strategies to be adopted in order to use preferentially renewable energy sources (RESs) to charge EVs
- Assessment of the EV potential to participate in the provision of power systems services, including reserves and power delivery, within a V2G concept

The replacement of conventional vehicles by EVs will also require specific local charging infrastructures. Several solutions may arise to fit different EV owners' needs:

- Charging stations dedicated to fleets of EVs
- Fast charging stations
- Battery swapping stations
- Domestic or public individual charging points for slower charging

Each of the above cases requires special consideration and as mentioned earlier, the stress each causes to the distribution system needs to be studied carefully. The distribution grid will obviously need reinforcement at several locations as per the strategy decided by the individual utility for EV charging.

7.6.5.1 PHEV characteristics [19]

To evaluate the impacts of PHEVs on the distribution system, it is first necessary to thoroughly explore the PHEV characteristics including battery capacity, state of charge (SOC), the amount of energy required for charging the battery, and charging level [19].

- The capacity of the PHEV battery depends on the type of vehicle and its driving range in the electrical mode designated as "all electric range (AER)."
- SOC is defined as the percentage of the charge remaining (stored) in the battery. It functions like fuel gauges in conventional internal combustion cars.
- The amount of energy required for charging PHEV depends on the energy consumed by the vehicle in daily trips, subsequently

Table 7.3 Size of Battery for Various PHEVs (kWh)

Type	Vehicle	PHEV30	PHEV40	PHEV60
1	Compact sedan	7.8	10.4	15.6
2	Mid-size sedan	9	12	18
3	Mid-size SUV	11.4	15.2	22.8
4	Full-size SUV	13.8	18.4	27.6

associated with daily miles driven and operation mode (i.e., electric motor or combustion engine).

- Charging level directly affects the duration time of charging such that a lower charging level increases the duration of charging. Generally at a slow charging rate, it takes 5 hours to reach the full charge when the battery is initially empty.

Thus, PHEV characteristics can be generally divided into two categories. The first category relates to those characteristics which are known based on the car manufacturer data or power system structure. The features such as battery capacity are placed in this category. The second class includes those properties that depend on traveling habits of the vehicle owner. Features such as daily miles driven and starting time of charging reside in this category. Thus, investigating owners' behavior, along with manufacturer data, is an important key to studying the effects of PHEV deployment in distribution systems.

Table 7.3 shows the battery capacity for PHEVs with all-electric ranges (AERs) of 30, 40, and 60 miles according to Reference [16]. Referring to Table 7.3, PHEVs have a wide range of battery sizes, from 7.8 to 27.6 kWh. Since the energy required for charging a PHEV depends on the battery capacity, it is necessary to determine the types and AERs of PHEVs to analyze their impacts.

7.6.5.2   PHEV impact on the grid [19]

There have been many studies conducted to evaluate the impact of PHEV penetration on the distribution grid. The peak load on the distribution system will increase substantially depending on the penetration of PHEVs. This is a big challenge. Even though the charging is planned at the off-peak hours at night, there is still a considerable increase in the peak load. There could be emergency charging during daytime as well. The increase in losses in the distribution system is also going to pose great problems for utilities. The loss curve of the utility will follow a similar pattern as the PHEV charging curve as per research. It sounds essential to control the time of charging PHEV to prevent the distribution congestion and feeder loss increment.

7.6.6 Microgrids [20,21,22,23,24,25]

As per the Microgrid Exchange Group of the Department of Energy (DOE), a microgrid is a group of interconnected loads and distributed energy resources within clearly defined electrical boundaries that act as a single controllable entity with respect to a grid. Microgrids thus comprise low-voltage (LV) distribution systems with distributed energy resources (DERs) (micro-turbines, fuel cells, photovoltaic generation, etc.) with storage devices (fly wheels, energy capacitors, and batteries) and flexible loads. Such a system can be operated in a nonautonomous way, if interconnected to a grid, or in an autonomous way, if disconnected from a main grid. The operation of micro sources in the network can provide distinct benefits to the overall system performance, if managed and coordinated efficiently [22,23].

The concept of microgrid thus applies to a local distribution system that encompasses generation, storage, and controllable loads. The generation will be a few kilowatts generally below megawatt (MW) level and can supply part or complete loads as required. The microgrid can operate in two modes, in the grid connected mode most of the time; however, there are times when the microgrid will operate in isolation mode, supplying partial or full load depending on the capacity of the DER. There are also isolated microgrids that operate without grid interfaces at remote locations where there is no access to grid supply.

A typical microgrid can be represented as in Figure 7.18.

Microgrid operation benefits can be listed as follows:

- Enables grid modernization
  - Serves as key component of grid modernization
  - Enables integration of multiple smart grid technologies
- Enhance the integration of distributed and renewable energy sources
  - Facilitates integration of combined heat and power (CHP)
  - Promotes energy efficiency and reduces losses by locating generation near demand

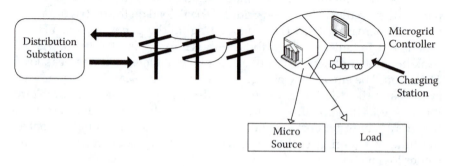

Figure 7.18 A typical microgrid.

- Helps reduce large capital investments by meeting increased consumption with locally generated power (local generation lowers investment in the macrogrid)
- Encourages third-party investment in the local grid and power supply
- Helps to reduce peak load
- Meets end user needs
  - Ensures energy supply for critical loads
  - Controls power quality and reliability at the local level
  - Promotes demand-side management and load leveling
  - Promotes community energy independence and allows for community involvement in electricity supply
  - Meets local needs and increases customer (end-user) participation
- Supports the main grid
  - Enables a more flexible macrogrid by handling sensitive loads and the variability of renewables locally
  - Enhances the integration of distributed and renewable energy resources including CHP
  - Supplies ancillary services to the bulk power system
  - Lowers the overall carbon footprint by maximizing clean local generation
  - Has the potential to resolve voltage regulation or overload issues

Microgrids are being created across the world owing to the following favorable factors:

- The cost of distributed generation is continuing to drop and is competitive with grid-supplied power in many regions.
  - Photovoltaic panels and inverters continue to decline in cost.
  - Clean natural gas-fired diesel generation is inexpensive because of very low gas prices.
- Environmental restrictions make older backup generation less attractive.
- Major storms have heightened the need for local backup power to withstand multiday outages, especially for loads that are critical to public safety, health, and welfare.
- The costs and complexities of microgrid design and grid interconnection are being addressed via numerous pilot projects.
- The penetration of grid interconnection of high renewables drives the need for more control of distributed resources and is another microgrid driver.
- Microgrids will help leverage or defer capital investment for energy and grid assets, as the microgrid encourages local generation.
- Microgrids enable innovation in new technology and services that have broad social impact beyond the local energy delivery.

Microgrids have been in existence at many university campuses and hospitals for a long time. Facilities like these were built around cogeneration plants that were capable of selling excess power to the grid at a wholesale level and were able to maintain electrical supply in the event of a grid outage. Combined heat and power facilities helped make the economics work. These facilities used waste heat from combustion turbines and/or combined cycle generators.

The IEEE smart grid global survey of smart grid executives cited three top benefits of microgrids: energy security and surety, renewable energy integration, and supply and load optimization. In addition, respondents said the top three industries likely to employ microgrids include hospitals, government, and utilities [22].

7.6.6.1   Types of microgrids

A microgrid can be set up and operated at a variety of scales, from a home (the smallest microgrid), to a low-voltage feeder microgrid, to a utility microgrid that owns a low-voltage grid. The classification is as follows [24]:

- *Private Industrial and Commercial Organizations*: These are privately owned and operated by facility managers and involve limited utility interactions. The primary focus is to support owners' industrial and commercial business operations with economic and reliable power supplies. Recently, we have seen the addition of college campuses—a new breed of microgrids with a focus on innovation.
- *Government Organizations*: Military base microgrids, for example, have a strong focus on energy reliability and safety. Government-owned microgrids often seek to improve economics by operating in parallel with utility grids. City and municipal microgrids usually function as drivers for a "smart city" vision.
- *Electric Utility Companies*: Vertically integrated utilities may deploy microgrids to serve customers with special, localized requirements. Deregulated utilities will collaborate with distributed energy resource aggregators to ensure service quality across the distribution grid and microgrids. Utilities may offer specialized expertise as a service to nonutility microgrid owners, as a means of increasing mutually beneficial interactions between the microgrid and the main grid. Some examples of the types of microgrid are given in future trends and practical implementation covered later in this chapter.

7.6.6.2   Microgrid control [26,27,28]

Microgrid control is a challenge, as it involves two modes of operation, grid connected and islanded, includes renewable resources, which deliver intermittent power, and pertains to distribution systems that have their own distinctive characteristics. In grid-connected microgrids, when there is an unbalance between the generation and the load, the power is instantly balanced by the inertia of the rotating system and the frequency changes. This is the basis of the primary control with active power and/or frequency drop control. However, since the islanded microgrid does not have enough spinning reserve, this kind of primary control is not inherently available. Also since microgrids are set up in the distribution system, the active power transfer mainly depends on the voltage magnitude, whereas in transmission systems, the active power transfer is mainly controlled by the phase-angle difference across the line. Another major concern in a microgrid control is the intermittency of the renewable sources that are integral parts of the microgrid, and this factor has to be accounted for when designing the microgrid controller. For the control in microgrids, new control concepts have been developed.

Three main control levels have been defined in such a hierarchy: primary, secondary, and tertiary. Figure 7.19 shows the control architecture of a microgrid, which consists of local and centralized controllers and communication systems.

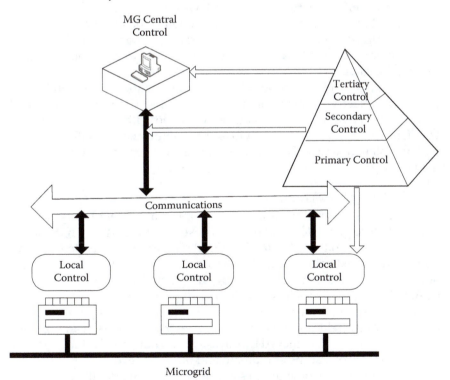

Figure 7.19 Microgrid control hierarchy [26],

The primary control ensures proper power sharing between the different DGs and local voltage control for stable operation of the microgrid. The primary control acts in milliseconds, due to the lack of inertia in the microgrid. For primary control, only local measurements are used, as in conventional grid control.

In grid-connected mode, with respect to the primary control, the DGs deliver power independent of the load variations. In the islanded mode, the DGs will deliver enough power as per the control strategy to ensure stable operation.

The secondary controller is a global controller with slower loops and low bandwidth communication to sense key parameters at certain points of the microgrid, generally a microgrid central controller (MGCC) and sends control commands to each DG unit. This is similar to the hierarchical SCADA control centers used in large transmission system frequency control. This centralized control for microgrids can also take care of the voltage control, power quality, and reactive power sharing in addition to the active power sharing. The centralized control can be decentralized by using new technologies such as multi-agent systems (MASs) for voltage and frequency control. The MAS allows intelligence to be distributed, where local controllers have their own autonomy and can make decisions. The centralized controllers consider the microgrid as a whole and optimize the operations, like an energy management system (EMS) in transmission SCADA.

Tertiary control is used for economic optimization, based on energy prices and electricity markets. The demand response mechanism and related tertiary control actions can be integrated with the secondary control. This centralized controller will exchange information with the distribution system operator (DSO) to optimize the microgrid operation within the utility grid.

7.6.6.3 DC microgrid [29]

DC microgrids interconnect a localized grouping of electricity sources and loads that generate, distribute, and use electrical power in its native DC form at low voltages (up to 1500 VDC) and are connected to the traditional centralized grid or function autonomously as physical and/or economic conditions dictate. Such microgrids are typically connected to and operate in conjunction with AC macro grids to form a smart grid [29].

DC microgrids have huge potential and hold the key in the electrification of commercial buildings in the future for many reasons. Huge savings in energy can be achieved by the use of DC power as it eliminates the AC-DC conversions at the equipment and building levels.

Most of the electronic devices are required for work, like computers, printers, smartphones, and smart lighting. The data centers with support IT systems all over the world use DC power. Electric vehicles, variable

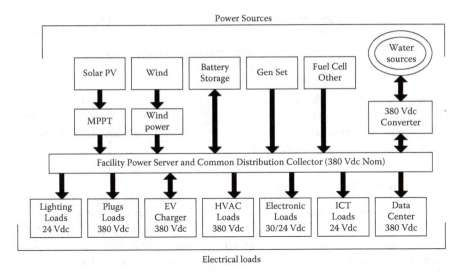

Figure 7.20 New microgrid power distribution topologies in buildings.

speed drives, and a large number of other devices use DC power. Generally, the AC supply is converted to DC power at the device level, and this leads to a waste of valuable energy. The solar panels produce DC power that is converted to AC power for distributing to the devices connected, and again converted to DC in the specific devices mentioned above.

Virginia Polytechnic Institute and State Unversity's Center for Power Electronics Systems in Blacksburg estimates that more than 80% of the electricity used in office buildings passes through power electronics and experiences one or more conversions between AC and DC electricity.

Figure 7.20 gives the framework for a DC microgrid with interconnected generation and loads at different voltage levels. The key application areas (for standardization of DC power use in buildings) include interiors and occupied spaces where lighting and control loads dominate the need for DC electricity data centers; telecom central offices with their DC-powered information and communications technology (ICT) equipment; outdoor electrical uses, including EV charging and outdoor light-emitting diode (LED) lighting; building services, utilities, and HVAC with variable-speed drive (VSD) and electronic DC motorized equipment.

The use of DC power is not without its challenges. These fall into five major categories:

- Lack of application and equipment standards for DC power distribution
- Lack of common understanding and basic application knowledge of building distribution-level DC
- Differences in safety and power protection device application

- Lack of a robust ecosystem to support the use of DC in building-level electrification
- An unclear pathway for moving from AC-centric power distribution to DC-inclusive distribution schemes

Attempts are being made to overcome these challenges, and the future of DC microgrid signals a paradigm shift in the centralized generation, transmission, and distribution of electricity to decentralized, local power generation in DC and consumption.

7.7   Smart transmission

Smart transmission involves the installation of phasor measurements units and developing wide area measurement systems and related applications, a complete description of which has already been provided in Sections 5.11 and 5.12. Integration of large renewables like wind and solar power also will have to be coordinated by the transmission control centers.

7.8   Lessons learned in deployment of smart grid technologies [30]

The smart grid deployments have taught stakeholders many lessons, some of which are discussed here.

Every day, our understanding of the smart grid as a "system of systems" evolves and improves and, therefore, so does the smart grid. Few experiences have helped foster our understanding more than the lessons learned through actual field deployments.

7.8.1   Lessons on technology

The challenge here is to separate hype created from the reality. The utility expectations are that the smart grid solutions are ready for implementation as a product, whereas the reality is that the technology is not that mature, and in many cases, the components were field re-engineered or upgraded to meet the objectives and expectations.

Integration and interoperability have been major technology issues, as smart grid deployment involves integrating products from multiple vendors. The lesson learned is to adopt and insist on standards and open architecture methodology to enable plug-and-play solutions.

Extensive laboratory testing for smart grid solutions is mandatory prior to implementation to understand the capabilities of the products and services offered, as redoing on site will be expensive and time consuming. Although individual components of the smart grid are thoroughly tested before installation, it has been discovered that some component technology may not be ready for integration into a solution. Often, these components do not interoperate well within the overall system, resulting in costly and time-consuming field re-engineering or upgrading. Each of the components of an integrated solution must perform according to its specifications. If any component underperforms, the entire solution underperforms. A chain is only as strong as its weakest link.

Take the case of monitoring the health of a critical bulk-power transmission transformer. The components include the sensors in or on the transformer, the monitoring and diagnostic equipment connected to the transformer, and two-way communications between the transformer and the maintenance office and the master station in the maintenance office. Bearing in mind that temperature variations affect the technology, when the monitoring and diagnostic equipment fail due to immature technology, the entire transformer monitoring solution fails.

Or consider phasor measurement units, phasor data concentrators (PDCs), two-way communications between the PDCs and the control center, and the wide-area measurement system (WAMS) in the control center. When the PDC fails because it cannot handle the amount and speed of synchrophasor data sent to it, the entire WAMS solution fails.

7.8.2 Lessons on implementation and deployment

The lesson learned is that integration and interoperability of each component in the system must be achieved to help ensure smart grid success. Testing an individual component is relatively easy, but extensive, end-to-end laboratory testing of components functioning within an operational system is mandatory prior to implementation to fully understand component capabilities and to help ensure interoperability before deployment as part of a larger solution. To develop an array of effective plug-and-play components, we must adopt—and insist upon—open standards and an open architecture methodology.

Compliance to standards does not guarantee interoperability. Coordinating software functionality with hardware from multiple suppliers has proved challenging. To ensure components successfully work together, they must comply with the same standard and have been tested for interoperability, especially if from different suppliers.

Imagine a utility with a SCADA/EMS (supervisory control and data acquisition/energy management system) from supplier X. This utility purchases an optimal power flow software application from supplier Y and wants to integrate it into supplier X's EMS. Typically, however, this is not possible because the EMS has a proprietary, real-time database structure. But if both suppliers have incorporated the common information model

(IEC 61968/61970) into their system and software application, the software application will successfully integrate with the system.

To take another example, suppose a utility is implementing volt-var control on its distribution feeders, where the logic resides in the substation. The substation controller communicates with the feeder-based intelligent capacitor bank controller using the DNP3 communications protocol (IEEE 1815). But the substation controller is from supplier X and the intelligent capacitor bank controller is from supplier Y. Though they both use the DNP3 communications protocol, they cannot talk with each other because of incompatibilities in the implementation of the protocol by both suppliers, which would have been identified and resolved if interoperability testing had been done. In this case, field re-engineering is needed to correct the incompatibilities.

Niche suppliers, though they provide valuable components and technologies, may have small engineering staffs that do not have the resources or familiarity to fully adopt and employ industry-wide standards, resulting in a lack of system interoperability.

Building long-term alliances with larger suppliers that have the resources to fully embrace industry-wide standards, while maintaining a holistic view of the overall solution, can help minimize interoperability issues. Larger suppliers also generally have engineering resources to provide field support, obviating the need to engage third-party field support; retaining third parties may open a can of worms in that they may not be familiar with the components or solutions that need re-engineering or upgrading. Rework, after all, is expensive and time consuming.

Packaged solutions from a defined group of strategically aligned suppliers will help improve coordination and interoperability of smart grid systems. These suppliers can work together to enhance equipment interoperability requirements, collaborate to resolve system problems, and develop documentation to improve personnel training.

7.8.3 Lessons on project management: Building a collaborative management team

Coordinating multiple suppliers as well as multiple internal departments within a utility—such as substation management, distribution engineering, and communications—has posed significant challenges. Collaboration is needed not only to develop technical standards but to effectively manage and steer smart grid projects as well.

Building an "A team" with the technical and project expertise to work collaboratively to identify and solve challenges is essential. Engaging a project manager with multidisciplinary authority for each smart grid solution can help utilities enhance departmental collaboration and interoperability efforts within the organization and when working with external vendors.

Establishing a program management office to oversee multiple project managers can help ensure adherence to overall program guidelines, including communications, status reporting, and risk management. Also, an interdisciplinary corporate steering committee—consisting of key stakeholders within the utility and within an alliance of suppliers—can essentially function as a "traffic cop" to direct project deployment in a controlled and timely manner while helping to mitigate risk.

7.8.4 Share lessons learned

Smart grid solutions involve a multitude of stakeholders, including residential and commercial customers, utilities, and strategic suppliers. Sharing information among all stakeholders is critical to success. An enormous amount of data is compiled every day from projects around the world, delivering new insights about equipment performance, system interoperability, new successes, and new challenges. Lessons need to be shared with all stakeholders so that data turn into knowledge that is actionable to help utilities, suppliers, and consumers build on past successes and avoid potential pitfalls. The development of "use cases" for each component and system is an effective means of disseminating lessons learned from deployments. Use cases can provide specific studies of how users interact with a system, besides giving a detailed description of a scenario. They can also define benefits, actors, functional requirements, business rules, and assumptions.

7.8.5 The lessons continue

The smart grid is a new, complex, and expansive system, and with each new project comes a new set of experiences and a new set of lessons to be learned. Adherence to industry standards and interoperability testing is critical for successful operations and performance success. As we continue to develop, test, and deploy smart grid solutions, we will continue to learn lessons that we can build upon to improve our performance and the performance of a smarter grid.

7.9 Case studies in smart grid

Smart grid implementation case studies are available all over the world, as the utilities have embraced smart grid in a big way. Here a few cases are presented which are typical in implementation and extension of SCADA systems to aid smart grid.

7.9.1   PG&E improves information visibility [31]

Pacific Gas and Electric (PG&E) is using PMUs to improve the distribution system, utilizing the 60 to 120 scans per second of the PMU data capture, for real-time view of the distribution system as a whole. This was done to help the operators anticipate and address the state and stability of the system. A distributed state estimation (DSE) platform was created, the processing of data was done in the field, and only relevant data were sent to the control center. The distribution substation-based state estimation for the utility was done using the selected real-time values. The implementation is underway as a pilot project in PG&E.

7.9.2   Present and future integration of diagnostic equipment monitoring [32]

The Omaha Public Power District (OPPD) has a number of equipment condition monitoring (ECM) devices that are monitored locally or by remote dial-in, and the data downloaded used to be stored in separate databases. The company found the usefulness of a corporate data warehouse to store all the information and use it more effectively for equipment diagnosis, as well as for substation training simulator usage. ECM was used to extend the life of a generator step-up (GSU) transformer with static electrification to enable OPPD to prepare a request for proposal (RFP) for a replacement transformer, evaluate the proposals, select a bidder, build and factory test the transformer, ship the transformer to the site, schedule an outage, and replace the transformer.

7.9.3   Accelerated deployment of smart grid technologies in India: Present scenario, challenges, and way forward [33]

As part of the smart grid vision, the Ministry of Power, Government of India, funded 14 smart grid projects to accelerate the power distribution sector reforms focusing on AMI, OMS, peak load management systems, renewable energy integration, and so on. The projects are underway and will be completed in 2015:

1. CESC (Karnataka)—AMI, outage management, peak load management, microgrid, and distributed generation with an initial 21,800 consumers in the Mysore Additional City area
2. Andhra Pradesh CPDCL—AMI, outage management, peak load management, and power quality management with 11,900 consumers in the Jeedimetla suburb of Hyderabad

3. Assam PDCL—AMI, outage management, peak load management, power quality management, and distributed generation with 15,000 consumers in the Guwahati area
4. Gujarat VCL—AMI, outage management, peak load management, and power quality management with 39,400 consumers in Naroda and Deesa
5. Maharashtra SEDCL—AMI and outage management with 25,600 consumers in Baramati in the Pune district
6. Haryana BVN—AMI and peak load management with 30,500 consumers in Panipat City
7. Tripura SECL—AMI and peak load management with 46,000 consumers in Agartala
8. Himachal Pradesh SEB—AMI, outage management, peak load management, and power quality management with 650 industrial consumers in Nahan
9. Puducherry electricity department—AMI with 87,000 consumers
10. JVVNL (Rajasthan)—AMI and peak load management with 2,600 consumers in Jaipur
11. Chattisgarh SPDCL—AMI with 500 industrial consumers in Siltara
12. Punjab SPCL—outage management with 9,000 consumers in Amritsar
13. Kerala SEB—AMI with 25,000 industrial consumers
14. West Bengal SEDCL—AMI and peak load management with 4,400 consumers in Siliguri town in the Darjeeling district

7.10   Summary

This chapter is an attempt to introduce smart grid concepts starting with the definition of a smart grid and moving on to a comparison of the old and the new electricity grids. A detailed discussion of the stakeholders in the smart grid development follows. The smart grid solutions are detailed thereafter where the discussion is about asset, demand, distribution, transmission, workforce and engineering optimizations, and smart meter and communications. Smart distribution components are discussed in detail, i.e., DER and energy storage, AMI, smart homes, PHEVs, and microgrids. Lessons learned in implementing a smart grid give a suitable conclusion to the chapter.

References

1. George W. Arnold, "Challenges and opportunities in smart grid: a position article," *Proceedings of the IEEE*, vol. 99, no. 6, June 2011, pp. 922–927.
2. Roger N. Anderson, Albert Boulanger, Warren B. Powell, and Warren Scott "Adaptive stochastic control for the smart grid," *Proceedings of the IEEE*, vol. 99, no. 6, June 2011.

3. James Momoh, "Smart grid: Fundamentals of design and analysis," Wiley-IEEE press, 2012.

4. Power and utilities changing workforce: Keeping the lights on.PWC, December 2012.

5. Peter Palensky, and Dietmar Dietrich, "Demand side management: demand response, intelligent energy systems, and smart loads," *IEEE Transactions on Industrial Informatics,* vol. 7, no. 3, pp. 381–388, August 2011.

6. Benjamin Kroposki, Pankaj K. Sen, Keith Malmedal, "Selection of Distribution Feeders for Implementing Distributed Generation and Renewable Energy Applications," *IEEE Transactions on Industrial Applications,* val. 49, no. 6, pp. 2825–2834, November/December 2013.

7. J.J. Iannucci, L. Cibulka, J.M. Eyer, and R.L. Pupp, Distributed Utility Associates Livermore, California, *"DER Benefits Analysis Studies: Final Report,"* NREL/SR-620-34636, September 2003.

8. Mauro Bosetti, Operation of Distributed networks with distributed Generation, Ph D Thesis, University of Bologna, 2009.

9. A. Mohd, E. Ortjohann, A. Schmelter, N.Hamsic, and D. Morton, "Challenges in integrating distributed energy storage systems into future smart grid," *in Proc. IEEE Int. Symp. Industrial Electronics,* June 30-July 2, 2008, pp. 1627–1632.

10. Grid 2030- A National Vision for Electricity's Second 100 Years, Based on the Results of the National Electric System Vision Meeting Washington DC, April2–3 2003.

11. IEC white paper on electrical energy storage, November 2011.

12. BAI Xiao-min, Meng Jun-xia, ZHU Ning-hui, "Functional Analysis of Advanced Metering Infrastructure in Smart Grid," *2010 International Conference on Power System Technology (POWERCON),* Oct. 2010, pp. 1–4.

13. Mohammad Ashiqur Rahman, and Ehab Al-Shaer, University of North Carolina at Charlotte, "AMI Analyzer: Security Analysis of AMI Configurations".

14. NETL Grid strategy, Powering our 21st Century Economy, Advanced metering infrastructure, US Department of Energy, Feb 2008

15. John D McDonald "Integrating DA With AMI May Be Rude Awakening for Some Utilities," *Renew Grid* (Feb. 20, 2013).

16. Hyunjeong Lee, Wan-Ki Park and Il Woo lee "A Home Energy Management System for Energy-Efficient Smart Homes," *2014 International Conference on Computational Science and Computational Intelligence,* vol. 2, March 2014, pp. 142–145.

17. B. Asare-Bediako, W.L. Kling, P.F. Ribeiro, "Home Energy Management Systems: Evolution, Trends and Frameworks," *47th International Universities Power Engineering Conference (UPEC),* 4–7 Sept. 2012, pp. 1–5.

18. J.A.P. Lopes, F.J.Soares, and P.M.R. Almeida, "Integration of electric vehicle-sin the electric power system,"*Proceedings of the IEEE,* vol. 99, no. 1, January 2011, pp. 168–183.

19. S. Shafiee, M. Fotuhi-Firuzabad, M. Rastegar, "Investigating the Impacts of Plug-in Hybrid Electric Vehicles on Power Distribution Systems," *IEEE Transactions on Smart Grid,* vol. 4, no. 3, pp. 1351–1360, Sept. 2013.

20. Nikos Hatziargyriou, "Microgrids: Architectures and Control,"Wiley-IEEE press, 2014. [Earlier it was referred as 23]

21. A. Papavasiliou, and S.S. Oren, "Large-scale integration of deferrable demand and renewable energy sources," *IEEE Transactions on Power Systems,* vol. 29, no. 1, January 2014.
22. Microgrids, large scale integration of micro-generation to low voltage grids. ENK5-CT-2002-00610, 2003-05.
23. Ralph Masiello and S.S. (Mani) Venkata, "Microgrids: There may be one in your *future,"IEEE Power and Energy Magazine,* vol. 11, no. 4, July/August 2103, pp.14–21.
24. Laurent Schmitt, Jayant Kumar, David Sun, Said Kayal and S.S. (Mani) Venkata, 'Ecocity Upon a hill', *IEEE Power and Energy Afagazine,* vol. 11, no. 4, July/August 2103, 59–70.
25. John D McDonald "Smart Grid Reality Check, Part III: Microgrids," *Renew Grid* (June 26,2013.
26. T.L. Vandoom, J.C. Vasquez, J. Dekooning, J. . Guerrero, and L. Vandevelde, "Microgrids-Hierarchical control and an overview of the control and reserve management strategies," *IEEE Industrial Electronics Magazine,* December 2013, pp. 42–55.
27. J.M. Guerrero, J.C. Vasquez, J. Matas, L. G. de Vicufia, and M. Castilla, "Hierarchical control of droop-controlled AC and DC microgrids- A general approach towards standardization," *IEEE Trans. Ind. Electron.,* vol. 58, no. 1, pp. 158–172, Jan. 2011.
28. J.M. Guerrero, M. Chandorkar, T.-L. Lee, and P. C. Loh, "Advanced control architectures for intelligent microgrids-Part I: Decentralized and hierarchical control," *IEEE Trans. Ind. Electron.,* vol. 60, no. 4, pp. 1254–1262, Apr. 2013.
29. Brian T. Patterson, "DC, come home", IEEE Power and Energy Magazine, vol. 9, no. 6, Nov-Dec 2012, pp 60–69.
30. J.D. McDonald, Lessons learned from field deployments. IEEE Smart Grid Newsletter, July 2011, www. Smartgrid.ieee.org/July2011.
31. Vahid Madani, Sakis Meliopoulos, PG & E Improves information visibility, T&D World, June 2014
32. M.I. Doghman, W.E. Dahl, John D McDonald, Present and future integration of diagnostic equipment monitoring, EPRI Substation Equipment Diagnostic Conference, 2002.
33. Datta, A., Mohanty, P., and Gujar, M., Accelerated deployment of Smart Grid technologies in India- Present scenario, challenges and way forward, IEEE ISGT, 2014, DOI: 10.1109/ISGT.2014.6816482

Further Readings

1. Stephen F. Bush, "Smart grid: Communication-enabled Intelligence for the electric power grid," Wiley-IEEE press, 2014.
2. Stuart Borlass, John D McDonald, "Smart grid: Infrastructure, technology and solutions," CRC press, 2012.
3. Mini. S. Thomas, Seema Arora, Vinay Chandna, "Distribution automation leading to a smarter grid," *IEEE ISGT Kallam,* India, Dec. 2011.
4. W.L.O. Fritz, "Smart grid - the next frontier," *Proceedings ofthe 9th Industrial and Commercial Use of Energy Conference (!CUE),* Aug.2012, pp. 1–3.

5. XinghuoYu, Carlo Cecati, Tharam Dillon, and M.G. Simoes, "The New frontier of smart grid," *IEEE industrial electronics magazine*, vol. 5, Sept. 2011, pp. 49–63.

6. Francisc Zavoda, "Advanced distribution automation (ADA) applications and power quality in smart grids," *China International Conference on Electricity Distribution*, Sept. 2010, pp. 1–7.

7. Richard E. Brown, "Impact of smart grid on distribution system design," *IEEE Power and Energv Society General Meeting - Conversion and Delivery of Electrical Energy in the 21st Century.* July 2008, pp. 1–4.

8. R. H. Lasseter, "Smart distribution: Coupled microgrids," *Proc. IEEE*, vol. 99, no. 6, June 2011, pp. 1074–1082.

9. R. W. Uluski, "The Role of advanced distribution automation in the smart grid,"*IEEE Power and Energv Society General Meeting.* July 2010, pp. 1–5.

10. B. P. Roberts and C. Sandberg, "The role of energy storage in development of smart grids," *Proceedings of the IEEE*, vol. 99, no. 6, June 2011.

11. C. Cecati, C. Citro, and P. Siano, "Combined operations of renewable energy systems and responsive demand in a smart grid," *IEEE Transactions on Sustainable Energy*, vol. 2, no. 4, October 2011.

12. Mini. S. Thomas, Seema Arora, Vinay Chandna, "Modeling of broadband power line (BPL) communication systems," *IEEE ISGT Kallam*, India, Dec. 2011.

13. S. Massoud Amin, Anthony M. Giaocmoni, "Smart Grid- Safe, Secure, self healing", *Power and Energy Magazine*, vol. 10, no. 1, Jan./Feb. 2012, pp. 33–40.

14. Julie Hull, Himanshu Khurana, Tom Markham and Kevin Staggs, "Staying in Control," Power and Energy Magazine, vol. 9, no. 1, Jan./Feb. 2012, pp. 41–48.

15. Mini. S. Thomas, Ikbal Ali and Nitin Gupta, " Interoperability framework for data exchange between legacy and advanced metering infrastructure," *International Journal of Energy Technology & Policies*, ISSN 2224-3232 (paper), vol. 2, no. 1, 2012.

16. *IEEE Standard for Synchrophasor Measurements for Power Systems*, IEEE Std. C37.118.1-2011 (Revision ofiEEE Std C37.118-2005), pp. 1–6,2011.

17. *IEEE Standard for Synchrophasor Data Transfer for Power Systems*, IEEE Std C37.118.2-2011 (Revision ofiEEE Std C37.118-2005), pp. 1–53,2011.

18. J. De La Ree, V. Centeno, J. S. Thorp and A. G. Phadke, "Synchronized phasor measurement applications in power systems," *IEEE Transactions on Smart Grid*, vol. 01, no. 1, pp. 20–27, June 2010.

19. Debomita Ghosh, T. Ghose and D. K. Mohanta, "Communication feasibility analysis for smart grid with phasor measurement units" *IEEETransactions on Industrial Informatics*, vol. 9, no. 3, pp. 1486–96, August 2013.

20. "Unified Real-time Dynamic State Measurements -A Report," POWERGRID, Feb'12.

21. C. Gouveia, Carlos Leal Moreira, J. A. Pecas Lopes, Diogo Varajao, And Rui Esteves Araujo, "Microgrid Service Restoration-The Role ofPlugged-In Electric Vehicles," *IEEE Industrial Electronics Magazine*, vol. 7, no. 4, December 2013, pp. 26–41.

22. N.Y. ISO, "Alternate Route: Electrifying the transportation sector, Tech. Rep., 2009.

Glossary

A

ACE, area control error
ADC, analog-to-digital converter
ADSS, all-dielectric self-supporting
AGC, automatic generation control
AMI, advanced metering infrastructure
AMR, automated meter reading
ANN, artificial neural network
AOR, area of responsibility
APCI, application protocol control information
APDU, application protocol data unit
ARP, Address Resolution Protocol
ASCII, American Standard Code for Information Interchange
ASDU, application size data unit
ASK, amplitude shift key

B

BPL, broadband over power lines

C

CDPD, cellular digital packet data
CFE, communication front end

checksum
CIS, customer information system
cordword
CRC, cyclic redundancy check
CRM, customer relations management
cross talk
CSMA/CD, carrier sense multiple access with collision detection
cybersecurity

D

DA/DMS, distribution automation/distribution management system
data mart
dataset
deadband
DFR, digital fault recorder
DISCO, distribution company
DLC, distribution line carrier
DNP3, Distributed Network Protocol version 3.3
DNS, Domain Name System
DPU, distributed processing unit
DSL, digital subscriber loop
DSO, distribution system operator
DTS, dispatcher training simulator

E

ECM, equipment condition monitoring
EISA, Energy and Independence Security Act
electromechanical
EMA, electric market authority
EMI, electromagnetic interference
EMS, energy management system
EPA, enhanced performance architecture
EPRI, Electric Power Research Institute
ESCO, energy service company
ESD, electrostatic discharge
Ethernet

F

FDIR, fault detection, isolation, and service restoration
FEP, front-end processor
fiber-optic

FSK, frequency shift key
FTIR, Fourier transform infrared
FTP, File Transfer Protocol

G

GENCO, generation company
GEO, geosynchronous Earth orbit
GIS, gas-insulated substation
GIS, geographical information system
GOOSE, generic object-oriented substation event
GOMFSE, generic object models for substation and feeder equipment
G&T, generation and transmission

H

half-life
HMI, human-machine interface
hydrothermal

I

ICCP, Inter-Control Center Protocol
ICMP, Internet Control Message Protocol
IEC, International Electrotechnical Commission
IED, intelligent electronic device
IGMP, Internet Group Message Protocol
I/O, input and output
ISO, independent systems operator
IVR, interactive voice response

L

LAN, local area network
LDC, load dispatch center
LEO, low Earth orbit
LFIS, load frequency isolation scheme
LPCI, link protocol control information
LPDU, link protocol data unit
LSB, least significant bit

M

MAC, media access control
MEO, medium Earth orbit

MGCC, microgrid central controller
MMS, manufacturing message specification
MU, merging unit
multicast
multiplex
multi-port communication
multi-protocol

N

NERC, North American Energy Reliability Corporation
NIC, network interface card

O

OFDM, orthogonal frequency division multiplexing
OLTC, on-load trap changer
O&M, operation and maintenance
OMS, outage management system
OPGW, optical power ground wire
OSI, open system interconnection

P

PCMR, protection, control, monitoring, or recording
PDC, phasor data concentrator
PLC, programmable logic controller
PLC, power line carrier
PLCC, power line carrier communication
PMU, phasor measurement unit
PQ, power quality
PSK, phase shift key

Q

QAM, quadrature amplitude modulation

R

RARP, Reverse Address Resolution Protocol
RFID, radio frequency identification
RIP, Routing Information Protocol
RISC, restricted instruction set computer
ROCOF, rate of change of frequency

RTO, regional transmission organization
RTU, remote terminal unit

S

SBO, select before operate
SCADA, supervisory control and data acquisition
smart grid
SCL, standardized configuration language
SCOPF, security constrained optimal power flow
SCTP, Stream Control Transmission Protocol
SMV, sampled measured value
SOE, sequence of events
switchyard
STLF, short-term load forecasting
STP, shielded twisted pair

T

T-101, T-102, T-103, T-104 (IEC)
TCP/IP, Transmission Control Protocol/Internet Protocol
TDM, time division multiplexing
TDMA, time division multiple axis
time stamping
TRANSCO, transmission company
TSDU, transport service data unit
TVE, total vector error

U

UART, universal asynchronous receiver transmitter
UDP, User Datagram Protocol
UHF, ultra-high frequency
UI, user interface
UPS, uninterrupted power supply
USART, universal synchronous/asynchronous receiver transmitter
UTP, unshielded twisted pair

V

VAR, volt-ampere reactive
volt-var
VSAT, very small aperture terminal

W

WAMS, wide-area monitoring system
WAN, wide area network
wastewater
waveform
WHAN, wireless home area network
WOC, wrapped optical cable

X

x-ray

Index

CPSIA information can be obtained
at www.ICGtesting.com
Printed in the USA
LVHW082102271221
707265LV00002B/171